Fluid Power
Educational
Series

Design of Pneumatic Systems

(In the SI Units)

Joji Parambath

Design of Pneumatic Systems
(In the SI Units)

Copyright © 2020 Joji Parambath

ISBN: 9798653408809

https://jojibooks .com

Disclaimer of Liability

The contents of this textbook have been checked for accuracy. Since deviations cannot be avoided entirely, we cannot guarantee full agreement. Only qualified personnel should be allowed to install and work on pneumatic equipment. Qualified persons are those authorised to commission, to ground, and tag circuits, equipment, and systems following established safety practices and standards.

Dedicated to

all my family members and friends in Palakkad

Table of Contents

Preface

A prerequisite for designing pneumatic systems is the knowledge of the functions, parameters, and specifications of the components needed for the power part, control part, and compressed air network of the system. At first, a preliminary design should be attempted as per the requirement specifications. The initial design must then be refined if required. Remember, the parameters of the system must synchronise with the data in the manufacturer's domain for the optimal design. Further, it is essential to incorporate inbuilt safety into the system.

The book explains the design aspects of pneumatic systems systematically. It also presents many typical examples of designing pneumatic systems, in the SI units, purely for educational or guidance purpose. The knowledge gained may be applied to develop more extensive industrial pneumatic systems.

Many other fluid power topics are given in other textbooks under the fluid power educational series by the same author. A list of all the books is given at the end of the book. Also, please see the details at https://jojibooks.com

Enjoy reading the book.
Your feedback is most welcome.

JOJI Parambath

Chapter 1 | Design Considerations

A pneumatic system must be designed to meet all the functional requirements of an application safely and efficiently. It must be economical and straightforward.

The system must provide the required performance, withstand operational hazards, and ensure its life expectancy. The design must also facilitate easy maintenance and the efficient removal of contaminants. Safety must be built into the system by incorporating interlocks, power-failure locks, and emergency shutdown features.

It must also take into account the speed of operation, the pressure and temperature ratings, the quality requirements of components, the cost of downtime and component replacement, the sensitivity to contamination, and the environmental conditions. The system must avoid component wear, overload, over-sizing, and high cost. It is also essential to prepare the circuit diagram of the system.

General Design Principles
Industrial pneumatic systems are designed with correctly-sized components and conductors. The use of undersized components and conductors can cause excessive pressure losses resulting from friction. As a consequence, the operating cost increases significantly. On the other hand, the use of oversized components and conductors impose higher capital and installation costs.

A simple and systematic approach is always a better way to design a pneumatic system optimally. Figure 1.1 outlines the critical steps involved in the design of a pneumatic system. These essential steps are: (1) System analysis and preparing specifications, (2) Circuit/Control system design, (3) Component selection and sizing, (4) Software simulation and analysis, (5) Development of system prototype, and (7) System performance evaluation and optimisation. The designer must take into account all working conditions specific to a given application. The following sections describe the essential steps for designing pneumatic systems.

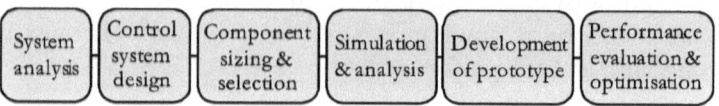

Figure 1.1 | Stages in the design process of a pneumatic system

System Analysis
The initial step in the design process of a pneumatic system is to define the exact requirements of the system. For understanding and defining the system requirements, it may be necessary to carry out detailed system analysis and develop operational specifications and system schematics.

It is essential to find the magnitude of each force (torque) and the type of motion required in the system. The type, size, and mounting styles of all system components must be investigated. The air consumption, stroke, duty cycle, and the required speed of actuators must also be examined.

The materials of construction of system components and the environment in which the system would be operated are other essential factors that must be considered while designing the system.

It is required to study many factors, such as load characteristics, actuator characteristics, sensor characteristics, and the extent of acceptable leakage while designing the system.

It may be required to familiarise with the technical parameters of solenoid valves and various electrical control components.

Factors, such as temperature, vibration, exposure to outdoor weather, moisture, initial costs, and maintenance costs, must also be considered while designing the system.

Many general requirements of the system, such as its robustness, compactness, quality, performance, efficiency, reliability, and safety, cannot be ignored.

The sequence of operations required in the application is to be clearly understood and detailed. It may also be necessary to construct the machine layout of the system.

Manufacturers' technical data must be consulted when configuring a solution to the design problem.

System Specifications and Circuit Design
Once the system requirements are established, system specifications and performance specifications must be prepared.

The next step is to develop suitable control circuits for the system using the symbols as per the relevant standard to meet the system specifications.

Compare any alternative circuit solutions concerning the power transmission efficiency and cost.

Component Selection
The first step in the selection of pneumatic components for the system is to outline the key factors that affect the selection of the components. For example, the targeted cycle time of work operations in an application is a critical factor when deciding the need for a double-acting cylinder.

Component sizing
Building the right pneumatic system for the specific application requirements is best achieved by first determining the parameters of the components of the system. The size or capacity of compressors, valves, actuators, conductors, and accessories to meet the performance specifications must be determined. It may be noted that manufacturers offer components with their sizes graded. Therefore, it is essential to synchronise the calculated

values of component parameters with the corresponding specifications of the components available in the market.

Simulation and Analysis
An appropriate software package can assist the designer in studying the interactions of the components of the system and analytically evaluating the performance of the system.

The software usually assists in component-level modelling as well as the system-level modelling and assessing their steady-state and dynamic responses.

Development of Prototype
Next, a prototype of the system must be developed to analyse, evaluate, and optimise the actual performance characteristics of the system.

Performance Evaluation
The performance of the newly-developed system must meet the required specifications, especially concerning its power, torque/force, speed, and efficiency, under the specified operating conditions.

Verify that the pressure and flow generated by the system meet the requirement specifications under all working conditions.

Verify that all the cylinders and motors in the system have enough strength to drive the associated loads and have enough capacity to accommodate side loads if any.

Ensure that the pressure losses, power losses, leakage, and heat generated in the system, under the worst operating conditions, are within limits.

If any shortcomings are observed, the system must be modified for its optimum performance.

Chapter 2 | Fundamentals of Pneumatics

Pneumatics is the branch of engineering sciences concerned with the transmission of energy using compressible fluids, like air, etc. Pneumatics is used throughout the industry due to the versatility and the simplicity of its application. Many characteristics make pneumatics more appropriate for many industrial applications than other types of power transmission systems.

Gas laws
The gaseous medium in a pneumatic system is sensitive to changes in volume, pressure, and temperature, and the gas laws govern its behaviour. Air is a mixture of gases and follows the laws of perfect gas concerning its behaviour in volumetric expansion or contraction and absorbing or releasing heat.

Boyle's law
Boyle's law gives the relation between pressure and volume of a gas. It states that 'at a constant temperature, the volume of a given mass of gas is inversely proportional to its absolute pressure.' Let V_1 (4 l) is the volume of a gas at pressure P_1 (1 bar) (See Figure 2.1). When the gas is compressed to a volume V_2 (2 l), then the pressure will rise to a value of P_2 (2 bar). Mathematically,

$$P_1V_1 = P_2V_2; \text{ where the temperature remains constant}$$

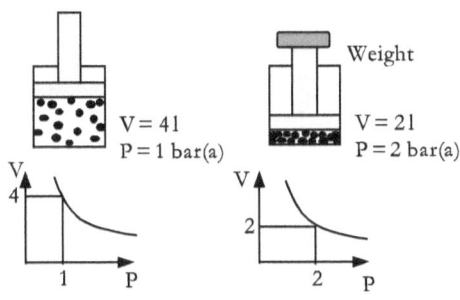

Figure 2.1 | Illustrating Boyle's law

Figure 2.1 illustrates the Boyle's law. The development of pressure, as the volume is reduced, is comparatively a slow process. This slow response necessitates the use of a receiver tank in a pneumatic system to store the compressed gas.

Gay- Lussac's Law

The law states that: 'If the volume of a given mass of gas is held constant, the absolute pressure of the gas varies directly to its absolute temperature.'

$$\frac{P1}{T1} = \frac{P2}{T2}; V \text{ remains constant}$$

Combined Gas Laws

The general law explains how the variables of absolute pressure, volume, and temperature are related to each other in a fixed mass of gas. The law can be expressed mathematically as:

$$P1 \cdot \frac{V1}{T1} = P2 \cdot \frac{V2}{T2}$$

Air Compression Process

The compression process of air can be thought of as taking place under isothermal or adiabatic or polytropic conditions.

Isothermal Compression

If the compression of air takes place under a constant temperature condition, the process is said to be isothermal. That means the heat of compression must be removed at the same rate as it is produced. Therefore, the process must be slow enough for the heat to dissipate from the air, as it is compressed. The equation governing the isothermal compression can be stated mathematically as:

PV is a constant

However, in practice, it is not possible to take out all the heat as it is generated.

Adiabatic Compression

When a volume of air in a system is compressed or expanded instantly, there is no time to add or dissipate heat into or out of the system, and this type of compression process is said to be adiabatic.

For example, the adiabatic compression takes place when air is compressed in a fully-insulated cylinder without any possibility of heat exchange with the surroundings. The same is the case with the air expanding through a nozzle very quickly.

The equation governing the adiabatic compression can be stated mathematically as:

$$P \, V^n \text{ is a constant}$$

The value of n for the adiabatic compression of air is taken as 1.4.

Polytropic Compression

An isothermal compression process must occur very slowly to keep the air temperature constant. An adiabatic compression process must occur very rapidly without any flow of energy into or out of the system. These compression processes are considered to be theoretical and hence are presumed to be taking place under ideal conditions.

In actual practice, compression of air occurs between the two limits of compression. A polytropic compression process represents the actual compression process in compressors operating under the normal rate of compression and expansion. For the polytropic compression:

$$P \, V^n \text{ is a constant}$$

The value of n for a poytropic compression depends on the rate of compression and is less than 1.4. Typically, for air n is taken as 1.3.

Characteristic Curves for the Compression Processes

The characteristic curves for the isothermal, adiabatic, and polytropic compression processes are given in Figure 2.2.

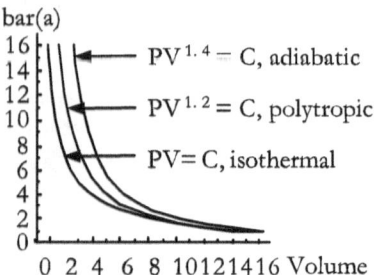

Figure 2.2 Characteristics of compression process for air

Pascal's Law

Pascal's law is central to the development of many fluid power devices, such as brakes, presses, and jacks. The law can be stated as follows: 'Pressure at any one point in a static fluid is the same in every direction' (Figure 2.3) and 'Pressure exerted on a confined fluid is transmitted equally in all directions, acting with equal force on equal areas'.

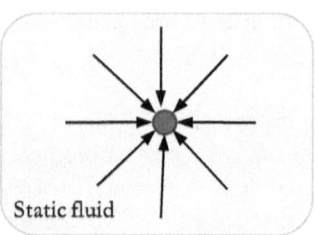

Figure 2.3 | Pascal's law

Pneumatic Pressure

Pressure in pneumatics operates according to Pascal's law. The pressure is the distributed response of force acting through a fluid.

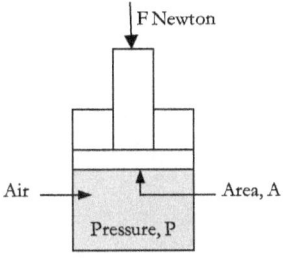

Figure 2.4 | Pressure development in confined air

In Figure 2.4, a definite amount of force (F) is applied to the air enclosed in the chamber through a piston of area A. The enclosed air is compressed, and its pressure (P) rises in direct proportion to the applied force and inverse proportion to the area of the piston. Pressure can, therefore, be defined as the force acting per unit area.

$$P = \frac{F}{A}$$

In the SI system, the unit of pressure is Pascal (Pa), and 1 Pascal is the constant pressure acting on a surface area of 1 square-metre with a perpendicular force of 1 Newton. For industrial pneumatic purposes, Pascal is too small a unit for use in measurements and hence more practical units like bar, kilo Pascal, and mega Pascal, are used.

1 Pascal	$= 1 \text{ N/m}^2$
1 bar	$= 100000 \text{ Pa} = 10^5 \text{ Pascal}$
1 Mega Pascal (MPa)	$= 10^6 \text{ Pascal} = 10 \text{ bar}$
1 Kilo Pascal (kPa)	$= 10^3 \text{ Pascal}$
1 bar	$= 0.1 \text{ MPa}$
1 bar	$= 14.5 \text{ Pound per square inch (psi)}$
1 bar	$= 1.02 \text{ kgf/cm}^2$
1 kgf/cm^2	$= 0.981 \text{ bar}$

Pressure scales

Everything on the earth's surface is subjected to a significant pressure head from the weight of the air above. This pressure is the 'atmosphere' (atm) and is approximately equal to 1 bar (1.01325 bar) at the sea level.

A pressure gauge is capable of measuring only the pressure concerning the local atmosphere. Therefore, the measured value of the pressure does not include the pressure exerted by the atmosphere. However, at times, we require pressure values with reference to the absolute vacuum.

According to the reference pressure levels, pressures in pneumatic systems can be specified in terms of the following two scales: (1) Gauge scale and (2) Absolute scale.

Gauge Scale

The gauge pressure is the pressure indicated by a pressure gauge installed at a particular location. It is the pressure above the local atmospheric pressure, regardless of the altitude. The gauge pressure, measured in bar, can be stated as bar(g) or simply bar.

Absolute Scale

The absolute pressure scale begins at the point where there is a complete vacuum (zero absolute pressure). Absolute pressure can be obtained by adding the datum pressure level (that is, 1.013 bar at sea level) to the gauge pressure. The absolute pressure, measured in bar, can be stated as bar(a).

Remember, absolute pressures are to be used in most of the calculations. Zero gauge pressure indicates the local atmospheric pressure (absolute).

The relationship between the absolute pressure and the gauge pressure is illustrated graphically in the figure. Figure 2.5 shows the pressure scales.

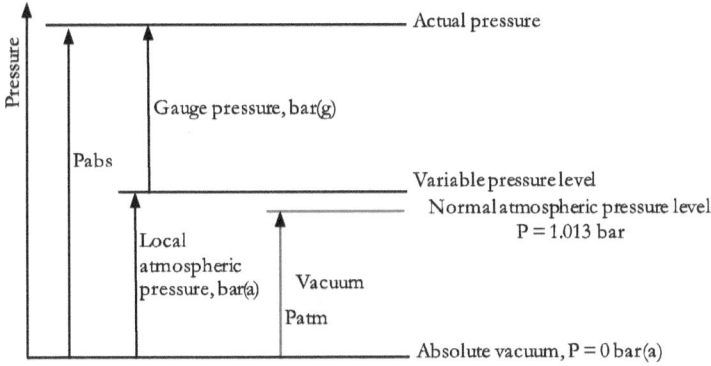

Figure 2.5 | Pressure scales

Economic Pressures in pneumatics

Pneumatic systems have been developed and progressed comparatively as a low-pressure system, as the compression of air is found to be a slow process. Pneumatic air consuming devices such as cylinders and air motors are generally rated for a maximum operating pressure of 8 to 10 bar.

However, practical experience has shown that 6 bar is the ideal pressure for the economic operation of the industrial pneumatic systems. This low pressure allows the designer to keep the size of pneumatic components very compact and maintain the cost of the components and piping system to a minimum.

Industrial Pressure Ranges

In most industrial pneumatic systems, the preferred operating pressure range is from 6 to 10 bar, as shown in Figure 2.6 Many popular air tools are engineered for pressures between 6 and 7 bar. However, the extended pressure range for industrial pneumatic systems can be up to 16 bar. Control pressures in pneumatics can be as low as 3 bar. The corresponding absolute pressure scale is also shown in the Figure.

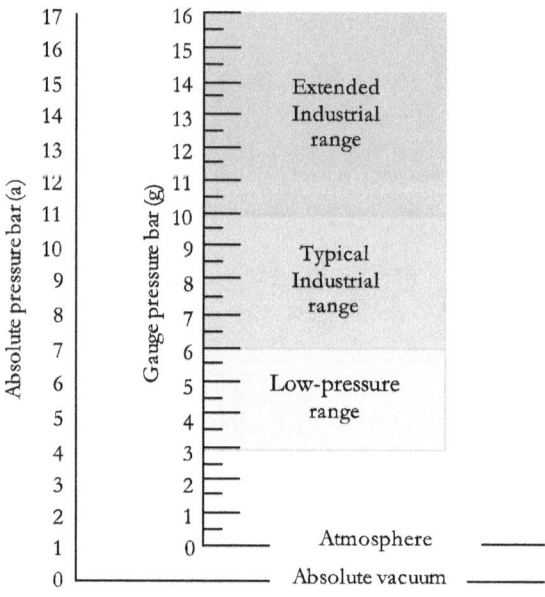

Figure 2.6 | Industrial pressure ranges

Problem 2.1

1120 litres of air at atmospheric pressure is compressed to 160 litres. What will be the gauge pressure developed, if the temperature remains the same?

Solution

Initial volume (V1)	= 1120 litre
Initial pressure (P1)	= 1 bar(a)
Final volume (V2)	= 160 litre
Final pressure (P2)	= V1 x P1 / V2
	= 1120 x 1 / 160
	= 7 bar(a)
	= 6 bar(g)

Pneumatic Force

Let us understand the process of developing a force to drive a load in the pneumatic system by the application of the pressure. Figure 2.7 shows the schematic diagram of a pneumatic cylinder with a piston. When the pressure (P) is applied to the area (A) of the piston, it develops a force (F). The amount of force developed is equal to the applied pressure times the area.

That is, $F = P \times A$

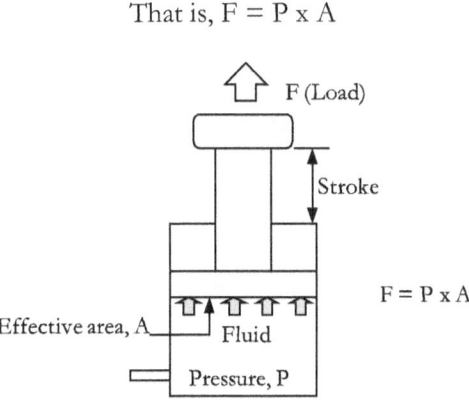

Figure 2.7 | Development of force (F) with the application of pressure (P)

Problem 2.2

Calculate the pressure produced by a force of 10000 N acting on the piston with an area of 0.0167 m².

Solution

Force	= 10000 N
Area	= 0.0167 m²
Pressure	= F/A
	=10000 / 0.0167 = 598800 Pa
	= 5.984 bar
	~ 6 bar

Force Multiplication

A pneumatic system can be designed for easy force multiplication. The basic idea of force multiplication is explained with the help of Figure 2.8. It shows an arrangement of two cylinders A and B with piston areas A1 (say, 1 cm²) and A2 (say, 10 cm²) (A2 > A1) respectively interconnected through a pipe. The enclosed space, inside the cylinders and the pipeline, is filled with air. When the cylinder A is applied with a force F1 (say, 10 N), a pressure P (10 N/cm²) is generated in the air medium. The same pressure P acts on the piston of the cylinder B, as per Pascal's law. This pressure causes the cylinder B to develop a force F2 (100 N). The governing equations for the forces developed in the cylinders are as follows:

$$F1 = P \times A1$$
$$F2 = P \times A2$$

Therefore,

$$F2 = F1 \times \frac{A2}{A1}$$

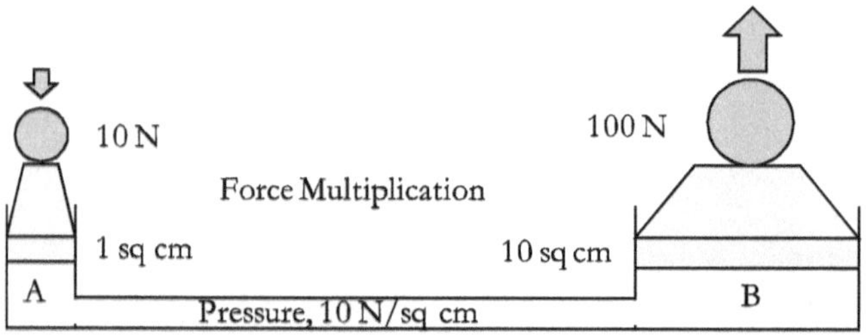

Figure 2.8 Illustration of force multiplication concept

Thus a pneumatic system can be designed for force multiplication. The ability of pneumatic systems to realise force multiplication can be thought of as leverage. However, it may be noted that the force multiplication is achieved by sacrificing distance. That is, for example, if the cylinder A moves by 10 cm, then the cylinder B moves by 1 cm.

The flow rate of Air

It is the volume of air passing a cross-section per unit of time under the specified conditions of pressure, temperature, and relative humidity. It is usually measured in terms of litres/minute (lpm) or cubic feet per minute (cfm).

Problem 2.3

What force is produced by a pneumatic cylinder with an area of 20 cm² operating at a pressure of 6 bar?

Solution

Area, A	$= 20 \text{ cm}^2 = 0.002 \text{ m}^2$
Pressure, P	$= 6 \text{ bar} = 6 \times 10^5 \text{ Pa}$
Force, F	$= P \times A$
	$= 6 \times 10^5 \times 0.002$
	$= 1200 \text{ N}$

Problem 2.3

A pneumatic lift arrangement consisting of a small cylinder of bore diameter 32 mm and a large cylinder of bore diameter 200 mm to lift a load of 4000 N. What is the force required to be exerted to the piston of the small cylinder to lift the load.

Solution

The bore diameter of the large cylinder = 200 mm
The bore diameter of the small cylinder = 32 mm
Force to be lifted, F2 = 4000 N

Piston area, large cylinder, A1 = $\prod$. 32²/ 4 = 803.84 mm²
Piston area, large cylinder, A2 = $\prod$. 200²/ 4 = 31400 mm²

Therefore,
Force needs exerted to cylinder A, F1 = F2 . (A1 / A2)
$$= 4000 \times (803.84 / 31400)$$
$$= 102.4 \text{ N}$$

Chapter 3 | Components of compressed Air Generation and Storage

The power source in a pneumatic system must be designed to supply a sufficient quantity of compressed air to all the system actuators for getting some useful work operations.

The primary functions of the power source include the generation and storage of compressed air, regulation of pressure, and removal of heat, solid contaminants, moisture, and oil from the compressed air.

Accordingly, the main components of the power source include a compressor, receiver tank, pressure regulators, aftercooler, filters, and dryer.

A compressor is designed to take in air at atmospheric pressure and deliver it into a closed system at a higher pressure for generating the forces needed to perform some useful tasks. The pressure in the system is regulated using pressure regulating units.

The compressed air preparation elements, such as aftercoolers, filters, and dryers, remove the undesirable elements, like heat, solid contaminants, moisture, and oil particles, present in the compressed air, in various stages.

Clean and dry compressed air is supplied to all the actuators in the system by a distribution system.

In short, the power source must be able to supply clean and dry compressed air at the required pressure and in sufficient quantity to all system actuators.

Air Compressors

Compressors are the most common industrial energy supply units. A compressor converts the mechanical power of its prime mover to pneumatic power using a compressible medium. It consists of a moving element enclosed in a housing. It is coupled to a diesel-operated, electric-operated, or gas-operated prime-mover. Electric models are the most popular for indoor applications.

However, it may be noted that the compression process in a compressor is slow. Therefore, a sufficient quantity of compressed air must be stored in a receiver tank.

Pressure Development in a Compressor

Figure 3.1 shows a reciprocating piston compressor connected to a reservoir and a table showing the pressure development with respect to time. It consists of a movable piston enclosed in a cylinder.

Assume that the compressor delivers 3 m³/min of air to the reservoir having a volume of 2 m³.

Using Boyle's law, the pressure rise in the prime-mover-driven compressor can be calculated easily, and the values of absolute and gauge pressures with respect to time are given in the table.

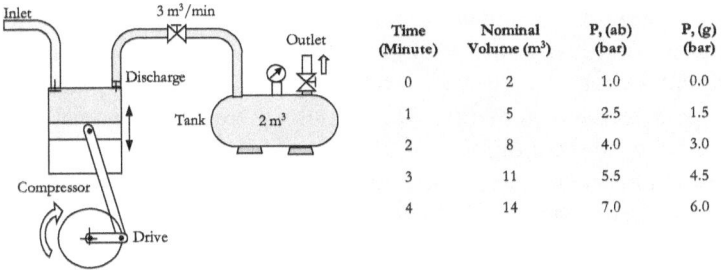

Time (Minute)	Nominal Volume (m³)	P, (ab) (bar)	P, (g) (bar)
0	2	1.0	0.0
1	5	2.5	1.5
2	8	4.0	3.0
3	11	5.5	4.5
4	14	7.0	6.0

Figure 3.1 | Pressure development in a compressor

Terms and Definitions, Compressor

A compressor is selected based on the volume of air it is required to deliver, the pressure it is required to operate at, and the quality of air that is needed. The factors about a compressor, which are most important to pneumatic personnel, are its working pressure, operating pressure, air delivery, the type of the drive unit, cooling methods and regulation. These terms are briefly explained below:

Working Pressure, Compressor

Working pressure is the pressure at the compressor outlet or in the receiver tank. This pressure is usually kept higher than that is required in the operating position.

Operating Pressure, Compressor

Operating pressure is the pressure that is required at the operating position.

Conditions of Air

The flow rate of compressed air in a pneumatic system is an essential parameter for sizing the components of the system. Remember, the conditions of air are different at different locations or under different situations. It is necessary to express them, under certain specified conditions of the air, to have an accurate representation of the flow rate.

The flow rate is generally defined in terms of the atmospheric conditions at a specified location. To be precise, the flow rate is defined in terms of a set of reference conditions of pressure, temperature, and humidity. Accordingly, there are many ways of specifying the conditions of the air. The important ways of representation are categorised under: (1) free air, (2) standard air, and (3) normal air.

Free Air

Free air is the air at the atmospheric conditions at the location of a compressor, but, unaffected by the compressor. This term does not mean air under standard conditions.

Air, under Standard Conditions

The standard set of reference conditions for representing the flow rate of air delivered by a compressor is specified in the ISO 1217 standard. This set of conditions is defined as the pressure of 1 bar(a), the temperature of 20^0 C, and relative humidity of 0%.

Air, under Normal Conditions

The normal inlet conditions of a compressor are specified as the pressure of 1.01325 bar(a), the temperature of 0° C, and Relative Humidity (RH) of 0%.

The summary of the conditions of air is given in Table 3.1.

Table 3.1 | Summary of conditions of air

	Pressure	Temperature	Humidity
Free air	Local conditions		
Standard air (ISO 1217)	1 bar(a)	20°C	0%
Normal air	1.01325 bar(a)	0°C	0%

The flow rate of Air

The flow rate of air, in respect of a compressor, is the volume of air displaced or delivered per unit of time at the rated speed of the driveshaft and under the rated conditions of pressure, temperature, and relative humidity. The flow rate can be measured in terms of: (1) Theoretical flow rate (or displacement) and (2) Effective flow rate (or delivery).

Theoretical flow rate (or Displacement Volume)

The theoretical flow rate is the quantity of inlet air that a compressor displaces. The theoretical flow rate of a compressor is the product of the volume of air swept in one revolution of its driveshaft and the number of revolutions per unit of time. It is usually expressed as litres per minute (lpm) or cubic feet per minute (cfm).

Also note: 1 cfm = 28.32 lpm and 1 lpm = 0.0353 cfm.

Effective flow rate (or Delivery Volume)
An effective flow rate is the quantity of air that a compressor delivers at the specified discharge pressure. The discharge pressure is typically specified at 6 bar. The quantity of the delivered compressed air is usually converted back to the actual inlet atmospheric conditions of the compressor at a given site or the standard (or normal) atmospheric conditions to normalise the effective flow rate. Accordingly, the effective delivery volume can be expressed in terms of the actual delivery volume (Free Air Delivery -FAD) or the standard (or normal) delivery volume.

Free Air Delivery (FAD)
It is the volume of compressed air delivered by a compressor at the specified discharge pressure (typically 6 bar) over a period and is usually stated in terms of the actual prevailing atmospheric inlet conditions. In other words, it is the expanded volume of air it forces into the associated system per unit of time. It is expressed in terms of lpm (fad) or cfm (fad).

Figure 3.2 shows the essential parameters required for the calculation of free air delivery. To calculate the free air delivery, firstly the atmospheric pressure (P1), actual temperature (T1), and humidity (RH1) at the inlet of the compressor are measured. Next, the maximum working pressure (P2), discharge temperature (T2), and the volume of the compressed air (V2) at the outlet are measured. Pv is the water vapour pressure. Finally, the volume V2 is referred back to the inlet conditions using the equation of the ideal gas. The value of V1 is the free air delivery of the compressor.

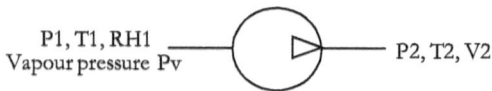

P1, T1, RH1
Vapour pressure Pv — P2, T2, V2

Figure 3.2 | Compressor parameters

$$\text{Free air delivery (FAD, V1)} = \frac{P2 \times V2 \times T1}{[P1 - (Pv \times RH1)] \times T2}$$

Standard (or Normal) delivery volume

It is the volume of compressed air delivered by an air compressor at the specified discharge pressure and generally stated in terms of the standard (or normal) atmospheric inlet conditions. It is expressed in terms of lpm (std or normal) or cfm (std or normal).

Problem 3.1

How much air under free air conditions (1 bar, 30°C) can be delivered by a receiver tank of 30 litres containing air at a pressure of 6 bar? The temperature inside the tank is 40°C. Neglect RH.

Solution

P1 = 6 bar = 7 bar(a) | P2 = 1 bar (a)
V1 = 30 litre
T1 = 40°C = 273+40 = 313 K
T2 = 30°C = 273+30 = 303 K

$$V2 = V1 \cdot \frac{P1}{P2} \cdot \frac{T2}{T1}$$

$$V2 = 30 \cdot \frac{7}{1} \cdot \frac{303}{313} = 203 \text{ litres (Free air)}$$

Problem 3.2

How much air under standard air conditions can be delivered by a receiver tank of 30 litres containing air at a pressure of 6 bar. The temperature inside the tank is 40°C. Neglect RH.

Solution

P1 = 6 bar = 7 bar(a) | P2 = 1 bar(a)
V1 = 30 litre
T1 = 40°C = 273+40 = 313 K
T2 = 20°C = 273+20 = 293 K

$$V2 = 30 \cdot \frac{7}{1} \cdot \frac{293}{313} = 197 \text{ litres (Std)}$$

Problem 3.3

How much air under normal air conditions can be delivered by a receiver tank of 30 litres containing air at a pressure of 6 bar. The temperature inside the tank is 40°C. Neglect RH.

Solution

$$P1 = 6 \text{ bar} = 7 \text{ bar(a)} \mid P2 = 1.01325 \text{ bar(a)}$$
$$V1 = 30 \text{ litre}$$
$$T1 = 40°C = 273+40 = 313 \text{ K}$$
$$T2 = 20°C = 273+0 = 273 \text{ K}$$

$$V2 = 30 \cdot \frac{7}{1.01325} \cdot \frac{273}{313} = 181 \text{ Nlitres}$$

Duty Cycle, Compressor

The duty cycle of an air compressor is the amount of time the compressor can run before it needs rest. For example, if the duty cycle of the compressor is 50% in one hour, then the compressor designed to run for a total of 30 minutes in one hour.

Classification of compressors

Compressors can be classified based on the design of their moving elements, the number of stages of the compression process, and the types of displacements. Compressors can be of lubricated, non-lubricated or oil-less designs. Oil is injected into the compression chamber of a lubricated compressor to lubricate its internal moving elements and bearings. It also takes away most of the heat produced in the compressor due to the compression process.

Reciprocating Vs Rotary Compressors

Compressors can be classified according to the specific design of the element used to create the flow of air. That is, reciprocating type or rotary type. Reciprocating piston compressors are very common and provide a wide range of pressures and delivery volumes. However, they are ideally suited for intermittent duty cycles.

Piston compressors are employed where high pressures (4 - 30 bar) and medium delivery volumes (< 18000 lpm) are needed. For higher pressures, multi-stage compressors with inter-cooling between each stage of compression are used. Rotary screw or rotary vane types are also used in many industrial applications. They are employed for applications with continuous duty cycles.

Classification: Single-stage Vs Multi-stage Compressors
Figure 3.3(a) shows a single-stage compressor. In a single-stage compressor, an increase in pressure takes place in a single cylinder. Single-stage compressors are generally used for pressures up to 12 bar. Single-stage compressors are adequate for small shops.

In a multi-stage compressor, as shown in Figure 3.3(b), the exhaust of one cylinder feeds the in-stroke of another in order to obtain higher outlet pressures. The multi-stage compressor is usually provided with an intercooler to remove the heat of compression. Double stage compressors can be used for getting pressures up to 30 bar. Multi-stage compressors are more efficient and have a longer service life than single-stage compressors. Two-stage compressors with oversized storage are required for large industrial operations.

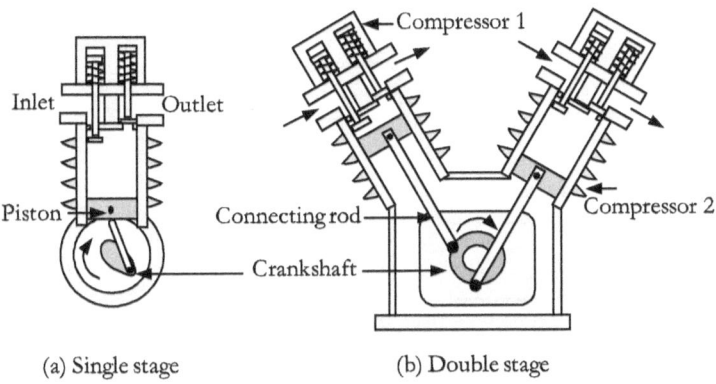

(a) Single stage   (b) Double stage

Figure 3.3 | Single-acting and double-acting compressors

Size Classification, Compressors

Compressor sizes range from a small compressor generating less than 20 lpm (FAD) with little or no preparation equipment to multiple compressor plant installations generating thousands of lpm.

Compressors with delivery volumes up to 2400 lpm and drive powers up to 15 kW are considered as small compressors. Compressors with delivery volumes in between 2400 to 18000 lpm and drive powers in between 15 to 100 kW are considered as medium-sized compressors. Compressors above the medium limit are considered as large compressors. The summary of size classification is given in Table 3.2.

Table 3.2 | Size classification of compressors

Size classification	Delivery volume lpm (FAD)	Drive power kW
Small compressor	<2400	<15
Medium-sized	2400 – 18000	15 – 100
Large compressor	>18000	>100

Positive Vs Dynamic Displacement Compressors

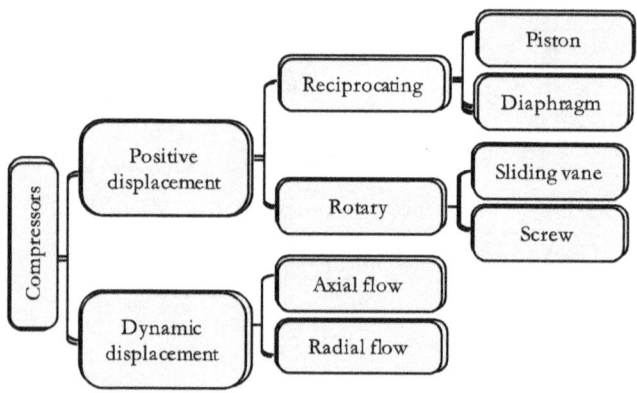

Figure 3.4 | Classification of compressors

A broad classification of compressors, according to the displacement, is shown in Figure 3.4.

In general, compressors are classified, according to the compressing elements used as: (1) positive displacement devices and (2) dynamic displacement devices.

In a positive displacement compressor system, the air is confined within an enclosed space where it is compressed by decreasing its volume.

In a dynamic displacement compressor, the air is accelerated by the rapidly rotating elements, such as rotor blades, causing some increase in pressure and a significant increase in velocity.

Reciprocating piston compressor

Figure 3.5 shows the basic single-cylinder reciprocating compressor. As the piston moves during its inlet stroke, the inlet valve opens and draws air into the cylinder. During the outstroke of the piston, the air is compressed and discharged through the outlet valve. Piston compressors have a relatively complex design with many moving parts.

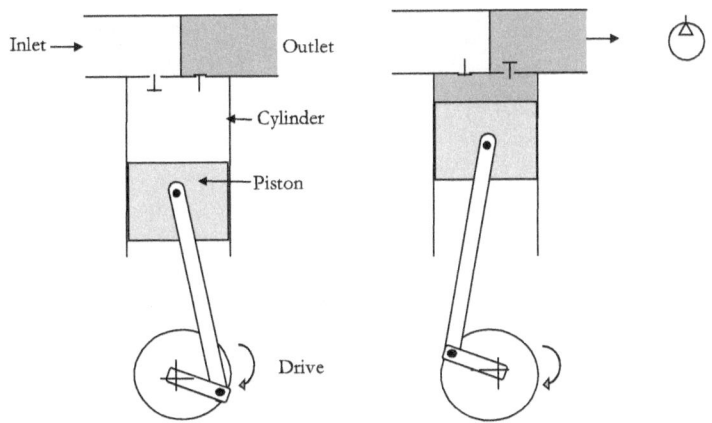

Figure 3.5 | Reciprocating compressor

Diaphragm compressor

In piston compressors, there is a likelihood of introducing small amounts of lubricating oil from the piston walls into the compressed air. This minimal oil contamination may be unwanted in food, pharmaceutical, and chemical industries, as well as in hospital and laboratory applications. For such applications, diaphragm compressors, as shown in Figure 3.6, may be used as the power source. A flexible diaphragm separates the compressor chamber and the actuating piston. This feature allows the lubricating oil to be excluded from the compressed air supply. Diaphragm compressors have limited delivery and pressure levels. They are used most often for light-duty applications.

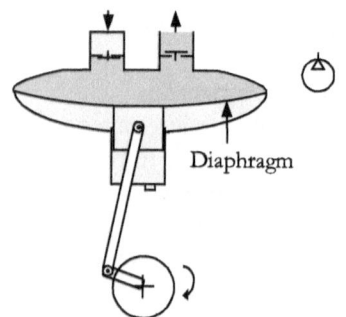

Figure 3.6 | Diaphragm compressor

Screw compressor

A screw compressor is shown in Figure 3.7. It consists of two helically grooved screws meshing with each other with a negligible clearance of about 0.05 mm. The design of the screws makes it possible to move air from the inlet to the outlet of the compressor. Compression is achieved by pushing the trapped air into a progressively smaller volume as the screws move ahead. Since there are no surfaces that make contact with one another, this type of compressor does not necessitate cooling and is characterised by the low noise level and a small loss of efficiency. They have the benefit of simplicity with fewer moving parts rotating at a constant speed and of a steady delivery of compressed air without pressure fluctuations.

Single-stage screw compressors are generally designed to operate at pressures less than 10 bar and capacities up to 6000 lpm. Higher pressures and capacities can be attained by multi-stage compression.

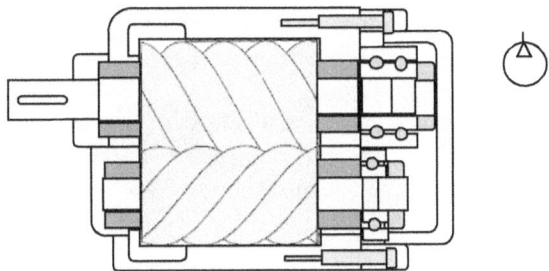

Figure 3.7 | Screw compressor

Vane Compressors

It consists of a prime-mover-driven rotor with sliding vanes in close-fitting radial slots, as shown in Figure 3.8. The rotor moves within a larger circular cavity. The centres of the rotor and the cavity are offset by a certain distance, causing an eccentricity. The vane tips bear against the casing and form an adequate seal. Side plates are used to keep the fluid confined to the space existing along the width of the rotor and vanes. Oil is injected into the compression chamber to act as a lubricant as well as a seal. It also removes the heat of compression.

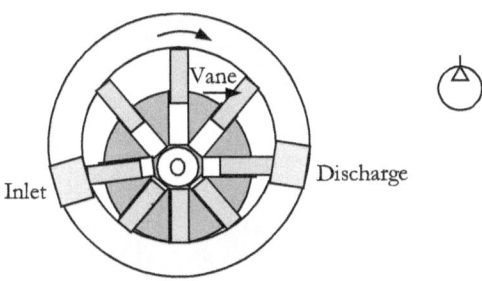

Figure 3.8 | Sliding vane compressor

As the rotor rotates, the space between any two successive vanes at the suction side increases. This expanding volume creates a partial vacuum, which draws air into the chambers formed by the vanes. The air is trapped in these chambers. The trapped air is then moved through the compressor by the rotating vanes. As the space between the two rotating vanes decreases, compressed air is squeezed out through the discharge port.

Sliding vane compressors are available with power ratings in the range from 7.5 to 150 kW and flow rate capacities from 1000 to 22000 lpm and discharge pressures from 5.5 to 9 bar.

Drive, compressor
Electrical motors or IC engines drive compressors. The electric drive can be single-phase or three-phase type. In factories, compressors are driven by 3-phase or single-phase induction motors. Power is transmitted through V-Belt, gear, or direct drive configurations.

Traditional V-Belt drives provide great flexibility in coupling the compressor and its prime mover at minimum cost. A gear drive provides a reduction in the axial load on the moving element of a compressor, thus extending its operational life. Gear drives typically use less energy than the V-belt drives. A direct drive can offer a compact configuration and minimum maintenance.

Belts and coupling used in the compressor system must be adequately shielded for safety.

Cooling of Compressors
Cooling fins on smaller compressors permit the heat to be removed by radiation. A large compressor is usually equipped with an additional fan to take away the heat by forced air cooling.

In the case of a compressor plant with a drive power in excess of 30 kW, the forced air-cooling is not adequate. The compressors are then installed with a water-circulation cooling system.

Storage of Compressed Air

An air receiver is a simple method of power storage. Figure 3.9 shows an air receiver with essential constructional features.

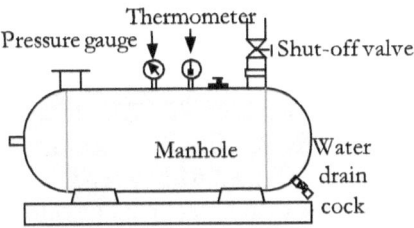

Figure 3.9 | Air receiver tank

Air receivers are available as horizontal models and vertical models. A vertical model is used when the floor space for its installation is limited. Air receivers must be provided with manual or automatic drain for releasing the collected water. They are usually made of mild steel.

The size of a compressed air receiver depends on the delivery volume of the compressor, load requirements, and the allowable pressure deviations in the receiver. Typical tank sizes (approx.) include: 5, 9, 10, 11, 12, 15, 17, 19, 23, 30, 34, 38, 57, 64, 75, 90, 98, 114, 225, 300, 450, 750, 900, 1500, 2500 litres.

Air receivers must be fabricated as per the relevant standards in one's region. Conformity to the standards ensures that the tank plates have sufficient thickness and are made with proper materials. During the fabrication, proper welding techniques must be applied by experienced operators.

An air receiver is also provided with: (1) a safety relief valve to guard against high pressures, (2) a pressure switch to sense the air pressure inside the tank, (3) a high-temperature switch to sense the excess air temperature inside the tank, and (4) pressure gauges for pressure indication.

Constructional Features, Compressor Unit

Compressors are available for light-duty or heavy-duty applications. They are typically constructed with a cast-iron cylinder and hardened steel crankshaft with precision bearings. Further, they are available as oil-free type or lubricated type.

A compressor can be of the open-frame type construction or the totally-enclosed type construction. The totally-enclosed type construction tends to reduce the compressor noise. Next, the drive part may also be provided with an enclosed belt guard. Air receivers are constructed with a steel tank in vertical or horizontal designs. They have a regulator, pressure switch, safety valve, fan-cooled motor, pressure gauge, and manual or automatic drain.

Typical Specifications, Compressor

The most important specifications for air compressors include the number of stages, compressed air delivery, maximum pressure rating, duty cycle, motor power, and operating voltage.

Some critical parameters of compressors for single-stage and multi-stage compressors are given in Table 3.3 and Table 3.4, respectively.

Single -Stage Compressors (Max Pressure 9 bar)

Table 3.3 | Single-Stage compressors (Max Pressure 9 bar)

Power	Drive speed	Displacement	FAD @6bar	Air Receiver
KW	rpm	lpm	lpm	litres
2.2	550	360	256	160
3.7	925	605	443	220
5.5	690	921	708	220
7.5	920	1228	950	220
7.5	920	1228	950	420
11	925	1853	1390	500
15	925	2406	1925	450
18.6	1050	3078	2463	450

Two-Stage Compressors (Max Pressure 12 bar)

Table 3.4 | Two-Stage compressors (Max Pressure 12 bar)

Power	Drive speed	Displacement	FAD @6bar	Air Receiver
KW	rpm	lpm	lpm	litres
2.2	925	311	250	160
2.2	925	311	250	220
3.7	925	501	410	220
5.5	1050	700	580	220
7.5	750	997	850	220
7.5	750	997	850	420
7.5	750	997	850	500
11	1150	1535	1250	500
15	1150	2195	1756	450
18.6	1150	2973	2379	450
22.4	1150	3363	2690	450

A rule of thumb suggests that on a steady pumping, a compressor will produce a minimum of 113 Nlpm flow of air for every HP capacity at 6 bar.

Pressure regulation
Pressure development in a compressed air system needs to be regulated to prevent the over-pressurisation of the system.

The pressure is regulated by using the most commonly used on-off regulation. As you are aware, pressure in a receiver is generally set higher than that is required at the operating position.

Safety Relief Valve
The receiver tank in a compressor is usually equipped with a safety relief valve to protect against any failure of its pressure regulation system.

Sizing an Air Compressor System

- Determine the compressor delivery and the tank size.
- To create an efficient compressed-air system, take into account the size of the compressor as well as the preparation and distribution of the compressed air. Sizing an air compressor requires a logical sequence of steps for selecting the right air compressor for specific applications.
- Check for the voltage and phase where the compressor is going to be located.
- Determine the compressor tank (receiver) size. Most manufacturers offer standard sizes based on the delivery volume of the compressor (the most popular sizes are 160, 220, 420, 500 litres

Compressor Selector Charts

Table 3.5, Table 3.6, and Table 3.7 can be used as a guide to select the kW rating of a compressor from the combined free air consumption of all pneumatic devices for applications in garages or industrial plants.

Table 3.5 | Compressor selector Chart for pressures ≤ 7 bar

Total air consumption, lpm		Pressure setting, bar		kW
Average use	Continuous operation	Lower limit	Upper limit	
Up to 187	Up to 50	5.5	6.5	0.37
190 – 300	50 – 85	5.5	6.5	0.56
300 – 385	85 – 110	5.5	6.5	0.745
385 – 575	110 – 165	5.5	6.5	1.12
575 – 750	165 – 215	5.5	6.5	1.5
750 – 920	215 – 290	5.5	6.5	2.2
920 – 1700	290 – 565	5.5	6.5	3.7
1700 – 2050	565 – 825	5.5	6.5	5.6
2050 – 2800	825 – 1130	5.5	6.5	7.45
2800 – 4250	1130 – 1700	5.5	6.5	11.2
4250 – 5650	1700 – 2265	5.5	6.5	15
5650 – 7100	2265 – 2800	5.5	6.5	18.6

Table 3.6 | Compressor selector Chart for pressures from 8 to 10 bar

| Total air consumption, lpm | | Pressure setting, bar | | kW |
Average use	Continuous operation	Lower limit	Upper limit	
Up to 100	Up to 30	8	10	0.37
100 – 200	30 – 60	8	10	0.56
200 – 280	60 – 80	8	10	0.745
280 – 425	80 – 120	8	10	1.12
425 – 700	120 – 210	8	10	1.5
700 – 1100	210 – 310	8	10	2.2
1100 – 1450	310 – 490	8	10	3.7
1450 – 1900	490 – 765	8	10	5.6
1900 – 2600	765 – 1050	8	10	7.45
2600 - 4000	1050 – 1600	8	10	11.2
4000 - 5400	1600 – 2100	8	10	15
5400 - 6800	2100 – 2750	8	10	18.6

Table 3.7 | Compressor selector Chart for pressures from 10 to 12 bar

| Total air consumption, lpm | | Pressure setting, bar | | kW |
Average use	Continuous operation	Lower limit	Upper limit	
Up to 330	Up to 100	10	12	0.745
330 – 525	100 – 150	10	12	1.12
525 – 685	150 – 170	10	12	1.5
685 – 1030	170 – 300	10	12	2.2
1030 – 1440	300 – 480	10	12	3.7
1440 – 1870	480 – 750	10	12	5.6
1870 – 2500	750 – 1000	10	12	7.45
2500 – 3825	1000 – 1560	10	12	11.2
3825 – 5240	1560 – 2125	10	12	15
5240 – 6650	2125 – 2700	10	12	18.6

Note: Data for different types of air compressors are given in Appendix 1

Sizing Air Receiver Tank using Formula

The size of the compressor tank can be determined based on the type of usage. If the usage is in short quick, concentrated bursts, then a small tank size can be used. If the unit is to sustain long periods of usage, a larger tank is required. The size of a receiver tank (V) can be determined using the following formula:

$$\text{Receiver size, V} = \frac{1.01 \times t \times (Q_r - Q_c)}{P_{max} - P_{min}}$$

Where,

V	= Size of the receiver tank, m^3
t	= Time to supply the required amount, min
Q_r	= Consumption rate of air, m^3/min (Std)
Q_c	= Compressor delivery rate, m^3/min (Std)
P_{max}	= Maximum pressure level in the receiver, bar
P_{min}	= Minimum pressure level in the receiver, bar

Problem 3.4

Calculate the minimum size of a tank that must supply air to a pneumatic system consuming 0.566 m^3/min (Std) for 6 minutes between 7 bar and 5.5 bar before the compressor resumes operation.

Solution

Q_r	= 0.566 m^3/min (Std)
Q_c	= 0
t	= 6 min
P_{max}	= 7 bar
P_{min}	= 5.5 bar

$$\text{Receiver size, V} = \frac{1.01 \times t \times (Q_r - Q_c)}{P_{max} - P_{min}}$$

$$\text{Receiver size, V} = \frac{1.01 \times 6 \times (0.566 - 0)}{7 - 5.5}$$

$$= 2.287 \ m^3 = 2287 \text{ litres}$$

Problem 3.5

Calculate the required size of a tank that must supply air to a pneumatic system consuming 0.566 m³/min (Std) for 6 minutes between 7 bar and 5.5 bar, if the compressor is running and delivering air at 0.1415 m³/min (Std).

Solution

$$Q_r \quad = 0.566 \text{ m}^3/\text{min (Std)}$$
$$Q_c \quad = 0.1415 \text{ m}^3/\text{min (Std)}$$
$$t \quad = 6 \text{ min}$$
$$P_{max} \quad = 7 \text{ bar} \mid P_{min} = 5.5 \text{ bar}$$

$$\text{Receiver size, V} = \frac{1.01 \text{ x t x } (Q_r - Q_c)}{P_{max} - P_{min}}$$

$$\text{Receiver size, V} = \frac{1.01 \text{ x 6 x } (0.566 - 0.1415)}{7 - 5.5}$$

$$= 1.715 \text{ m}^3 = 1715 \text{ litres}$$

Sizing Air Receiver Tank using Charts

An air receiver tank can also be sized by referring to charts. The sizing based on the airflow capacity is given in Table 3.8, and the sizing based on the compressor power is given in Table 3.9.

Table 3.8 | Receiver tank size Vs airflow capacity

Airflow capacity, lpm	Receiver tank size, litre
283	37.85
566	75.70
850	113.56
1133	151.41
1416	189.27
2124	283.90
2832	378.50
4248	567.81
5664	757.08

Table 3.9 | Receiver tank size Vs compressor power

Compressor power, kW	Receiver tank size, litre
3.7	75.70
5.6	113.56
7.5	151.41
11.2	227.12
14.9	302.83
18.7	378.50
22.4	454.25
29.8	605.66
37.3	757.08
44.8	908.50
56	1135.62
74.6	1514.16

Air Compressor Packaged Units
Most industrial air compressors are supplied as self-contained packages.

An air compressor packaged unit is fully assembled compact air compressor system, complete with an air compressor, receiver tank, inlet filter, electric motor, belt/gear/direct drive, and automatic microprocessor controllers.

An electronic controller is provided for the intelligent shutdown of the unit and energy-saving operation.

Next, optional equipment includes an aftercooler, particulate filters, an integrated dryer, an automatic moisture drain, a low oil safety control, a magnetic starter, a cooling fan, and a pressure reducing valve.

The packaged units are available with low noise enclosure (65 - 70 dBA) and vibration isolators for the quiet operation.

Chapter 4 | Components for the Removal of Contaminants in Compressed air

Contaminants can enter a pneumatic system through the air taken in by the system compressor. In industrial surroundings, air carries a large number of solid impurities and moisture. Contaminants are harmful to pneumatic systems.

Solid Contaminants
Solid impurities are industrial dust which includes iron, carbon, silicates, fibre-glass, soot, and other abrasive materials.

Humidity
Humidity is the moisture present in the atmosphere. Moisture is in the form of water vapour that remains suspended in the given volume of air. It is difficult to remove the moisture from the air. Humidity is usually expressed in terms of either absolute humidity or relative humidity.

Absolute Humidity
Absolute humidity is the actual amount of moisture present in one cubic metre of air.

For example, as shown in Figure 4.1(a), if 8.7 grams of moisture is present in one cubic metre of air at a particular temperature, say at 20°C, then its absolute humidity is 8.7 g/m³ at 20°C. It is always temperature-dependent.

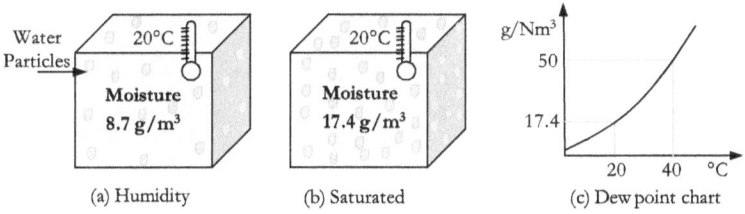

Figure 4.1 | Absolute humidity concept

Saturation Quantity

A given volume of air at a specified temperature can contain moisture in the vapour form up to its saturation level.

For example, as shown in the dew point chart of Figure 4.1(b), the following can be observed. At 20°C, one cubic meter of air can contain a maximum of 17.4 gm of moisture. At 40°C, it can contain a maximum of 50 gm of moisture.

The saturation quantity is a function of temperature and is given by the dew point chart [See Figure 4.1(c)].

Relative Humidity

Relative humidity (RH) of air is the ratio of its absolute humidity to the air saturation quantity at a given temperature. It is usually expressed as a percentage. That is,

For example, if one cubic-metre of air contains 8.7 grams of moisture at 20°C, then the relative humidity can be calculated as follows:

Saturation quantity at 20°C (From Dew point chart) = 17.4 gm

$$RH = \frac{\text{Absolute humidity}}{\text{Saturation quantity}} \times 100\% = \frac{8.7}{17.4} \times 100 = 50\%$$

It may be noted that 100% RH means the given volume of air is saturated. The relative humidity is dependent on both temperature and pressure.

Decreasing the temperature or increasing the pressure will result in condensation of excess moisture above the saturation level.

Oil Particles

Oil is used as a lubricating or working medium in many industrial machines. Therefore, in industrial surroundings, air carries harmful oil particles that can be far more than 10 mg/m³.

Air Quality Classification

ISO 8573-1: 2010 stipulates contaminants and quality classes of compressed air for general use. Air contains solid, water and oil particles as contaminants. The standard specifies the amount of contamination allowed in each cubic metre of compressed air. A quality (or purity) class number is defined for each of these contaminants according to the permissible levels of specific parameters. These parameters and permissible values of them against each class are given in Table 4.1.

Table 4.1 | Permissible levels of contaminants (ISO 8573-1)

Class	Solid particulates Max. particles per m³ 0.1 – 0.5 micron	0.5 – 1 micron	1 – 5 micron	Mass concentration, mg/m³	Water Vapour pressure dew point	Liquid water g/m³	Oil mg/m³
0	As per a written specification between user and supplier (more stringent than class 1)						
1	≤20000	≤400	≤10	-	≤-70°C	-	0.01
2	≤400000	≤6000	≤100	-	≤-40°C	-	0.1
3	-	≤90000	≤1000	-	≤-20°C	-	1
4	-	-	≤10000	-	≤+3°C	-	5
5	-	-	≤100000	-	≤+7°C	-	-
6	-	-	-	≤5	≤10°C	-	-
7	-	-	-	5-10	-	≤0.5	-
8	-	-	-	-	-	0.5-5	-
9	-	-	-	-	-	5-10	-
x	-	-	-	>10	-	>10	>10

An air quality class is specified as a combination of the three air quality numbers. For example, a quality class 1.2.1 means that in each m³ of compressed air, the particulate count should not exceed 20000 particles in the 0.1-0.5 µ size range, 400 particles in the 0.5-1 µ size range and 10 particles in the 1-5 µ size range. A pressure dewpoint of -40°C or better is required, and no liquid water is allowed. In each cubic metre of compressed air, not more than 0.01mg of oil is allowed. This weight of the oil is the total level for liquid oil, oil aerosol and oil vapour.

Preparation of Compressed Air

The compressed air delivered by a compressor has many harmful contaminants and objectionable conditions. The contaminants and conditions are enumerated below:

- The compressed air is very hot.
- It contains a very high concentration of solid particles
- It contains moisture in the vapour and liquid forms
- It contains oil in the liquid and vapour forms

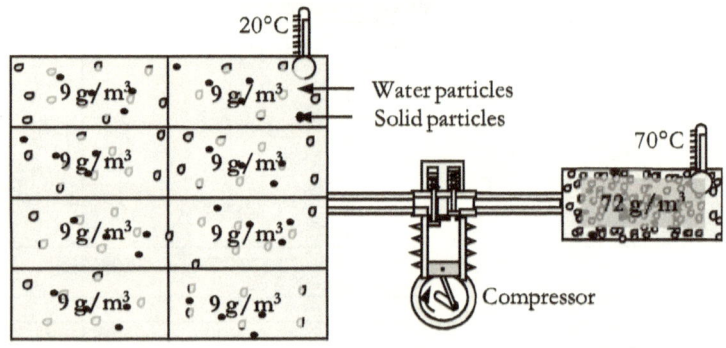

Figure 4.2 | Condition of air at the outlet of a compressor

The condition of the compressed air at the outlet of the compressor is shown in Figure 4.2.

Effects of Contamination

Solid contaminants can damage compressor seals. They can disturb the operation of the sophisticated downstream components such as valves and actuators.

Water droplets resulting from the condensation can cause rusting of exposed surfaces, the formation of sticky emulsions, and consequent jamming of valves.

Stages of Compressed Air Preparation

To achieve any degree of reliability, the components of pneumatic systems must get clean and dry air. Hence, air must be prepared or conditioned before it can be allowed to go into a pneumatic system. Preparation of compressed air consists of reducing its temperature, removing water and solids from it, regulating its pressure, and in many cases introducing lubricant in it. In general, the preparation of air falls into three distinct stages, as shown in Figure 4.3.

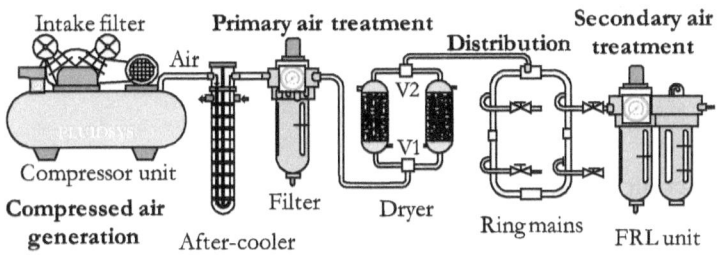

Figure 4.3 | Stages of compressed air preparation

Intake Filter

The intake filter removes large particles, usually of sizes greater than 200 microns, which can damage the air compressor.

Compressor and Storage Unit

A compressor generates compressed air. A receiver tank stores the compressed air.

Primary Air Treatment

Primary air treatment is intended to reduce the temperature of the air at the outlet of the compressor, remove solid contaminants usually larger than 100 microns, present in it and dry the air to reduce its humidity.

The units used in the primary air treatment are aftercooler, mainline filter, and dryer.

Aftercooler

Typically the temperature of air exiting a compressor is in between 82°C to 177°C. An aftercooler is intended to reduce the temperature of compressed air discharged by a compressor to approximately 2.7 or 5.5 or 8.3 or 11°C over the temperature of the cooling medium.

A stand-alone aftercooler is a separate unit installed downstream of the compressor. Compressor manufacturers may include aftercoolers within the compressor package.

Types of Aftercoolers

The two basic types of aftercoolers are: (1) Air-cooled aftercoolers and (2) Water-cooled aftercoolers.

Air-cooled Aftercooler

Air-cooled aftercoolers use ambient air to cool the hot compressed air. Figure 4.4 gives the essential parts of an air-cooled aftercooler. The compressed air travels through finned tubes while ambient air is forced over the cooler by a motor-driven fan. The forced air removes heat from the compressed air. As the air cools, the moisture in the compressed air condenses. When the air reaches the cooler's separator, centrifugal motion causes the condensed water and other contaminants to hit the cylinder walls and drip down to the drain.

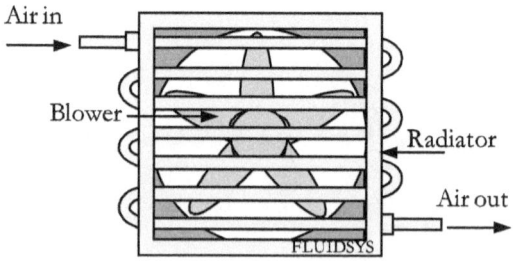

Figure 4.4 | Air-cooled aftercooler

Water-cooled Aftercooler

The standard style is the shell and tube aftercooler in which a bundle of copper tubes fitted inside the shell, as shown in Figure 4.5. The hot compressed air flows through the tubes in one direction while the cooling water flows in the opposite direction around the tubes in the shell. As the compressed air is cooled, moisture would condense out of the air in the form of water. The moisture separator and drain valve remove the water.

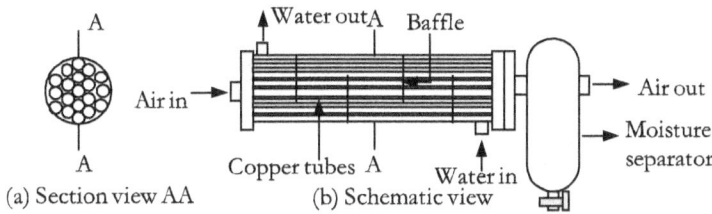

Figure 4.5 | Water-cooled aftercooler

The tube bundles can be fixed or removable. Fixed tube bundles cost less but are more difficult to maintain as compared to removable bundles that can be detached for cleaning or service.

Condensate Separator, Aftercooler

It will remove the liquid condensate. It is installed at the base of the aftercooler.

Automatic Condensate Drains, Aftercooler

Condensate drips down the walls of the separator into the automatic condensate drain. As the drain fills, a float rises, opening a valve that empties the condensate. This automatic action ensures that condensate and other contaminants do not build up in the cooler.

Table 4.2 gives typical specifications for air-cooled aftercoolers, and Table 4.3 gives typical specifications for water-cooled aftercoolers.

Typical Specifications of Air-cooled Aftercoolers

Table 4.2 | Specifications of air-cooled aftercoolers

Capacity lpm	Max working pressure bar	Max air inlet temp °C	Approach temperature (°C)	Electric supply (VAC)	In/out ports BSP
967	16	120	$t_{amb} + 11°C$	220	½"
1384	16	120	$t_{amb} + 11°C$	220	¾"
2834	16	120	$t_{amb} + 11°C$	220	1½"
4251	16	120	$t_{amb} + 11°C$	220	1½"
5668	16	120	$t_{amb} + 11°C$	220	2"NB flg
8502	16	120	$t_{amb} + 11°C$	220	2"NB flg
11336	16	120	$t_{amb} + 11°C$	415	2"NB flg
17003	16	120	$t_{amb} + 11°C$	415	2"NB flg
21254	16	120	$t_{amb} + 11°C$	415	2"NB flg
28339	16	120	$t_{amb} + 11°C$	415	2"NB flg

t_{amb} = Ambient temperature, °C

Typical Specifications of Water-cooled Aftercoolers

Table 4.3 | Specifications of water-cooled aftercoolers

Capacity lpm	Max working pressure bar	Max air inlet temp °C	Approach temperature (°C)	Air in/out ports BSP	Water in/out ports BSP
1384	16	150	$t_{amb} + 11°C$	3/4"	1/2"
2834	16	150	$t_{amb} + 11°C$	1"	1/2"
5583	16	150	$t_{amb} + 11°C$	1½"	3/4"
7000	16	150	$t_{amb} + 11°C$	2"NB flg	1"
11166	16	150	$t_{amb} + 11°C$	3"NB flg	1½"
13916	16	150	$t_{amb} + 11°C$	3"NB flg	1½"
20833	16	150	$t_{amb} + 11°C$	3"NB flg	2"
27833	16	150	$t_{amb} + 11°C$	4"NB flg	2"
35000	16	150	$t_{amb} + 11°C$	4"NB flg	2"
41666	16	150	$t_{amb} + 11°C$	5"NB flg	2"

t_{amb} = Ambient temperature, °C

Compressed Air Filters

Filters can be fitted to the mainline of a compressed air system to remove dust, dirt, oil, and water from the compressed air. They can be used to clean the air to a recognised compressed air purity standard, such as ISO 8573. They are often used with refrigerant and desiccant type dryers. A schematic diagram of a pneumatic mainline filter is shown in Figure 4.6.

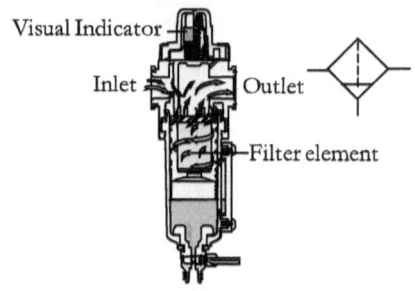

Figure 4.6 | Mainline filter

Types of Filters

Pneumatic filters can generally be classified as: (1) General purpose filters, (2) Coalescing filters, and (3) Adsorbing filters.

General-purpose filters can remove solid particles down to 5 microns and water droplets from the system.

Coalescing filters can remove 99.99% of oil contaminants and solids greater than 0.3 microns in size, in general. It can provide air quality 1.7.2 as per the ISO standard 8573-1. However, it cannot remove oil vapours.

An adsorbing filter is an ultra-high efficient coalescing filter with active carbon pack. The filter media can attract and remove oil vapours, and the carbon pack can remove hydrocarbon gases. It can provide air quality 1.7.1 as per the ISO standard 8573-1.

Typical Specifications of Filters

Table 4.4, Table 4.5, and Table 4.6 give typical specifications for general-purpose filters, coalescing filters, and adsorbing filters, respectively. [1 l/s = 2.11888 scfm]

Table 4.4 | Sample specifications of general-purpose filters

Capacity (l/s)	Working pressure (bar)	Element size (μm)	Connection size (BSP)
33	16	5, 25, 40	G¼
66	16	5, 25, 40	G⅜
75	16	5, 25, 40	G½
75	16	5, 25, 40	G¾
160	16	5, 25, 40	G¾
190	16	5, 25, 40	G1
200	16	5, 25, 40	G1¼
200	16	5, 25, 40	G1½

Table 4.5 | Sample specifications of coalescing filters

Capacity (l/s)	Working pressure (bar)	Element size (μm)	Connection size (BSP)
16	16	0.01	G¼
28	16	0.01	G⅜
28	16	0.01	G½
28	16	0.01	G¾
35	16	0.01	G½
35	16	0.01	G¾
60	16	0.01	G1

Table 4.6 | Sample specifications of adsorbing filters

Capacity (l/s)	Working pressure (bar)	Element size (μm)	Connection size (BSP)
7	16	0.01	G¼
11	16	0.01	G⅜
11	16	0.01	G½
11	16	0.01	G¾
25	16	0.01	G½
35	16	0.01	G¾
60	16	0.01	G1

Drain

A drain mechanism can be provided in filters to remove the water before it re-enters into the downstream air. The mechanism can be of the manual type or automatic type. The standard drain is a manual type. The auto drain facility ensures that the filter bowl will be drained without operator intervention. There are two types of automatic drains. They are: (1) Float type and (2) Differential pressure type.

Float type drain works on the float principle. As water accumulates in the bowl, the float will be lifted. As a result, a passage will open in the bowl. The air pressure in the bowl then will vent to the atmosphere through the opening, blowing the water away from the bowl. When the water is completely drained, the float drops back into the orifice, sealing off the passage in the bowl. If an application is continuous, a float type drain must be used in a filter used in the application.

In the differential pressure design, there must be a pressure differential across the filter.

An external drain is used where severe condensation problem exists. The water will be removed with the self-flushing action of the filter.

Additional Specifications of Filters

Filter bowl:
- Transparent (Polycarbonate) with guard
- Metal with liquid level indicator

Service life indicator: Mechanical | Electrical

Drain: Manual | Auto drain | External

Threads: PTF | ISO taper | ISO parallel

Dryers

The natural water vapour content of air is concentrated and is carried through the compression process as a vapour in high temperatures. For simple applications, all that may be essential is an aftercooler, air receiver, and filter with condensate traps to remove the excess humidity.

Additional means of dehydration must be provided where the demand for high quality compressed air is entailed. The most commonly used methods of compressed air drying are: (1) Adsorption drying and (2) Refrigeration drying.

It is worthwhile to use an aftercooler before any air dryer to reduce the amount of work enforced on the dryer. An air dryer is ideally fitted downstream of the compressor and reservoir.

Adsorption dryer

Adsorption is the physical process of collection of moisture on the porous surface of certain granular materials such as silicon dioxide (Silica gel), activated alumina, and copper sulphate. Figure 4.7 shows the constructional features of a typical adsorption dryer. When compressed air is passed through the drying agent, the moisture present in the air is adsorbed by the drying agent until it gets saturated. Dry compressed air is delivered out of the dryer.

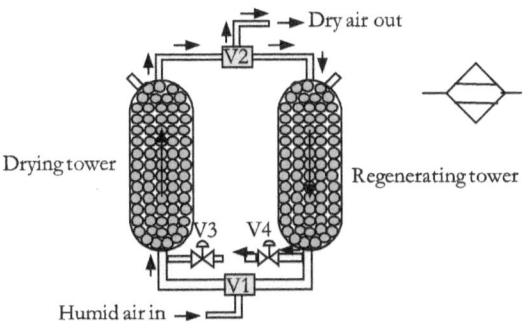

Figure 4.7 | Adsorption dryer

The silica gel drying agent changes its colour as it approaches the saturation point. When saturated, the drying agent can be renewed by blowing warm or cold air through the material, which then takes up the moisture. The silica gel reassumes its original colour when the moisture is driven out. This method of drying is also known as regenerative drying.

In practice, two parallel chambers are used for non-stop production. While one chamber is drying the air, the other one can be set for regeneration.

Types of Adsorption Dryers
Adsorption dryers can be of the following three types: (1) Heatless type, (2) Heated type, and (3) Heated blower type.

- Heatless twin tower dryer diverts a portion of the dried air to the off-line tower. This dry air then flows through the saturated desiccant and regenerates it.

- In the heated type dryer, a portion of the dried air is first passed through a high-efficiency external heater before entering the off-line tower to regenerate the desiccant.

- Heated blower type dryer employs a high-performance centrifugal blower to direct ambient air through a heater and then through the off-line tower. The stream of heated air then regenerates the desiccant.

Refrigerated dryer
The schematic of a typical refrigerated air dryer is shown in Figure 4.8. It consists of a heat exchanger and a refrigerating unit. In the first stage, the warm and humid compressed air is passed through the heat exchanger. The air gets cooled to the near ambient temperature condition. The moisture present in the air gets condensed corresponding to the temperature and water is precipitated.

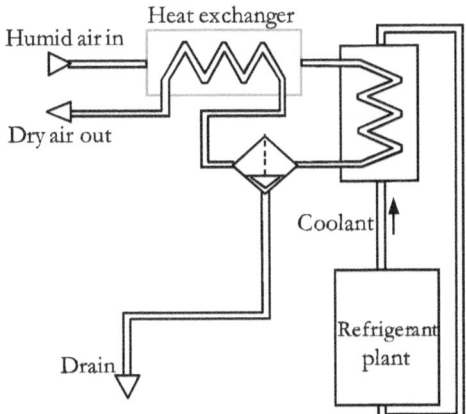

Figure 4.8 | Low-temperature dryer

In the second stage, the partly prepared air is passed through the refrigerating unit to reduce the temperature of the compressed air to as low as 2°C. Again the moisture is condensed corresponding to the temperature in the refrigerating unit. The condensed water can then be collected in the water traps provided at appropriate points. Finally, the air once again goes through the heat exchanger and gets discharged in a clean and dry condition.

Pressure Dew Point
The pressure dew point is the lowest air temperature reached during the drying process at the specified operating pressure. For example, the typical pressure dew point that can be achieved with the adsorption dryer is about −40°C, and that of the refrigerated dryer is about 2°C.

Selection of Dryers
The selection of dryers depends on the variables, such as system demand, compressed air capacity, air quality requirements and applicable life cycle costs that are unique to the compressed air system.

Table 4.7 to Table 4.11 give sample specifications for various types of dryers.

Typical Specifications of Basic Desiccant Dryers

Table 4.7 | Typical specifications of basic desiccant type dryers

Capacity (Nm³/min)	Pressure dew point, °C	Pressure rating (bar)	Connection (in)
2.5	-40	10	1 NPT
3.4	-40	10	1 NPT
4.5	-40	10	1½ NPT
5.7	-40	10	1½ NPT
7.1	-40	10	1½ NPT
8.5	-40	10	2 NPT
11.3	-40	10	2 NPT
14.2	-40	10	2 NPT
17	-40	10	2 NPT
22.7	-40	10	3 NPT
28.3	-40	10	3 NPT
34.0	-40	10	3 NPT
42.5	-40	10	4 FLG
51.0	-40	10	4 FLG
59.5	-40	10	4 FLG
76.4	-40	10	4 FLG
93.4	-40	10	6 FLG
113.3	-40	10	6 FLG
141.6	-40	10	6 FLG

Typical Specifications of heated type Desiccant Dryers

Table 4.8 | Typical specifications heated type desiccant dryers

Capacity Nm³/min	Pressure dew point, °C	Heater rating of dryer, if used kW	Pressure rating bar	Connection in
4.2	-40	2	10	1 NPT
5.7	-40	3	10	1½ NPT
7.1	-40	3	10	1½ NPT
8.5	-40	3	10	1½ NPT
11.3	-40	4.5	10	2 NPT
14.2	-40	4.5	10	2 NPT
17	-40	6	10	3 NPT
22.7	-40	9	10	3 NPT
28.3	-40	9	10	3 NPT
34.0	-40	12	10	3 NPT
42.5	-40	15	10	3 NPT
51.0	-40	18	10	4 FLG
59.5	-40	18	10	4 FLG
84.9	-40	30	10	4 FLG
113.3	-40	36	10	6 FLG
141.6	-40	50	10	6 FLG
169.9	-40	60	10	6 FLG
226.5	-40	75	10	8 FLG

Typical Specifications of heated Blower type Desiccant Dryers

Table 4.9 | Typical specifications heated blower type desiccant dryers

Capacity Nm³/min	Heater rating of dryer, if used, kW	Blower rating of blower, if used, kW	Connection (in)
4.2	3	0.75	1 NPT
5.7	4.5	0.75	1½ NPT
7.1	6	1.1	1½ NPT
8.5	6	1.1	1½ NPT
11.3	9	1.5	2 NPT
14.2	12	1.5	2 NPT
17	12	3.7	3 NPT
22.7	18	3.7	3 NPT
28.3	24	5.6	3 NPT
34.0	24	5.6	3 NPT
42.5	30	11.2	3 NPT
51.0	36	11.2	4 FLG
59.5	45	11.2	4 FLG
84.9	60	14.9	6 FLG
113.3	80	18.7	6 FLG
141.6	100	22.4	6 FLG
169.9	125	22.4	6 FLG
226.5	175	29.8	8 FLG

Pressure dew point -- 40°C | Pressure rating – 10 bar

Typical Specifications of Refrigeration Dryers

Table 4.10 | Specifications of refrigeration dryers
(230V, 1Ph, 50 Hz)

Capacity (lpm)chk	Electrical power, kW	Pressure rating (bar)	Connection (in)
0.6	0.29	3 to 16	G ¾
0.85	0.29	3 to 16	G ¾
1.25	0.28	3 to 16	G ¾
2.1	0.55	3 to 16	G 1
2.55	0.64	3 to 16	G 1
3.2	0.76	3 to 16	G 1¼
3.9	0.95	3 to 16	G 1¼
4.7	1.13	3 to 16	G 1¼
5.65	0.86	3 to 16	G 1½
7.0	1.1	3 to 16	G 1½
8.25	1.4	3 to 16	G 2

Table 4.11 | Specifications of refrigeration dryers
(400V, 3Ph, 50Hz)

Capacity (lpm)	Electrical power, kW	Pressure rating (bar)	Connection (in)
11.5	1.08	7	G 2
12.5	1.12	7	G 2
14.5	1.48	7	G 2
17.0	1.39	7	DN 65
23.0	1.94	7	DN 80
28.0	2.43	7	DN 80
34.0	2.72	7	DN 80
45	3.28	7	DN 100
52	3.89	7	DN 100
65	4.83	7	DN 150
78	5.88	7	DN 150
98	9.82	7	DN 150

Pressure dew point +3°C | Inlet air temperature +60°C

Chapter 5 | Components for the compressed Air Distribution System

The objective of the air distribution system is to act as a leak-proof carrier of compressed air and limit pressure drops within permissible limits. The air distribution system is made up of conductors and fittings, which interconnect various components of a pneumatic system. A typical pneumatic distribution layout is shown in Figure 5.1. A well-organised industrial pneumatic distribution system is designed with correctly sized pipes and components, ensuring a minimum number of elbows and bends so that pressure energy is not unnecessarily wasted. Distribution of compressed air should be planned and executed carefully by taking into account the following considerations: (1) correct sizing of pipes and fittings, (2) choice of pipe materials, (3) pipe layout, and (4) the total cost of the conductor system.

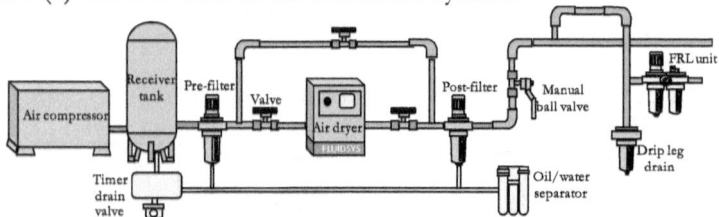

Figure 5.1 | A typical pneumatic layout

Conductors
The conductors are generally divided into three classes: (1) Pipe (Rigid), (2) Tubing (Rigid or semi-rigid), and (3) Hose (Flexible). More than one type of conductor may be used in the same installation.

Rigid pipe
The main distribution system is made up of rigid pipelines, feeder lines, and associated fittings and accessories. Copper, iron, steel, and aluminium pipes must be brazed and welded or can be joined using threaded connectors. Welded connections are robust and

leak-free and are the primary choice for fixed main distribution pipelines. As a rough rule, piping is employed for diameters above 50 mm. Air hoses and tubing may be employed for conducting compressed air to air-powered tools and equipment, instruments and gauges.

Tubing
There are semi-rigid type and flexible type of tubing available for use in pneumatic applications.

Examples of the rigid type of tubing are steel, aluminium, copper and polyvinyl chloride, and flexible type are nylon and polyethylene. Nylon tubes are robust and can be used for a variety of applications within general pneumatics. Polyurethane tubes are extra flexible and soft and are especially suitable in applications where the requirement of short bending radii for tubing is indispensable.

Each material has definite characteristics, which make it more appropriate for some services than others. Since tubing can be bent, lines from tubing require only a minimum number of fittings. A tube is usually specified by outside diameter and wall thickness.

Plastic tubing is flexible and has gained full acceptance in industry for use as conductors in pneumatic systems as it is inexpensive and extremely easy to use with a high degree of flexibility. Food grade tubes are colourless and tasteless and will not pass on extraneous flavour or odour to susceptible foods or beverages.

Hose
Hose assemblies are used to connect compressed air source to actuators that must be located on movable parts or because of the necessity to bend lines. The advantages of using hoses are that it can be easily installed, requires fewer installation skills than that required for pipes or rigid tubing, is capable of absorbing shock, and is readily available in a whole range of pressure ratings.

Flexible hoses are manufactured from natural and synthetic rubbers and several plastics, and they are reinforced by fabric or wire braid. A few examples are: (1) polyester-reinforced PVC hose and (2) metal braided rubber hose. A hose is usually specified by the inside and outside diameters. A hose should have a smooth bore and must be resistant to oil vapours and lubricants. The wall of the hose must be sufficiently hard to resist heavy impacts, and shock blows. The outer structure of the hose must be strong and abrasive resistant.

Fittings

Pipes and tubes are joined to other pipes and tubes or the components of an installation by using some connectors. Remember, there are many different types of connectors available for pneumatic systems. Some examples of fittings are push-in fittings, push-on fittings, and compression fitting. Push-in fittings are used for quick and straightforward assembly of pneumatic circuits. They are very compact units comprising retained collets and positive tube anchorage for easy tube insertion and hence for rapid assembly.

Fittings are made of materials such as stainless steel, aluminium, bronze, and plastic with silicon-free nitrile rubber / Viton 'O' rings. They are available in a variety of shapes to form unions, elbows, tees, nipples, caps, plugs, couplings, and crosses.

Quick-disconnect coupling

Quick-disconnect couplings are widely used in pneumatic systems, mainly where there are frequent needs to uncouple the lines for maintenance, testing, and safety. Many disconnect couplings have double checks that can be used for easy detachment without any loss of compressed air.

Air Fuse

Tubing and hose which can cause damage through whiplash when severed have to be protected using an air fuse.

Pipe threads

Threaded pipes connections must contain male threads on the pipes. Threads are available to a variety of standards, and some of which are: American National Pipe Threads (NPT), Unified Pipe Threads (UNF), British Standard Pipe Threads (BSP), and Metric Pipe Threads (M). The choice between these standards is determined by the standards already chosen for a user's region or country. Taper threads are cone-shaped and form a seal between the male and female parts as they tighten, with assistance from some jointing compound or plastic tapes.

Flow resistance

The flow of compressed air through piping creates friction and consequent pressure drop. It may be noted that the pressure loss is proportional to the square of the velocity of the flow. Elbows, T-pieces, two-way valves, and slide valves are also responsible for the interference with the flow and the corresponding loss of pressure. However, this pressure drop cannot be avoided but can be considerably reduced by routing pipes properly and assembling the fittings correctly.

Pipe layout

Various piping arrangements can be used in air distribution systems depending on usage requirements, size of the plant and delivery volume. Generally, distribution is arranged as a manifold, as shown in Figure 5.2(a) or as a ring main, as shown in Figure 5.2(b).

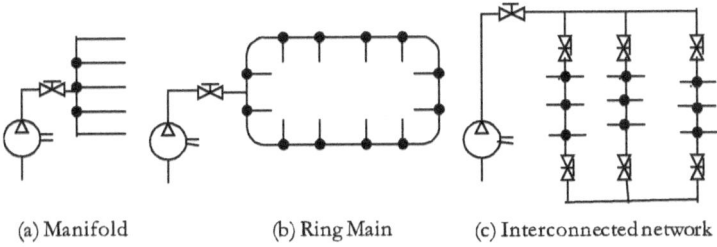

| (a) Manifold | (b) Ring Main | (c) Interconnected network |

Figure 5.2 | Pipe layouts

As the actuating devices consume air, the pressure is decreased at the downstream. One technique of compensating the pressure drop is to use the ring-main layout. Any demand for the compressed air can be met in two directions using this layout. A ring main ensures largely uniform pressure conditions in the air network.

With an interconnected network system as shown in Figure 5.2(c), parts of the ring can be separated using the shut-off valves for maintenance, repair, and extension of the network without disturbing the rest of the system.

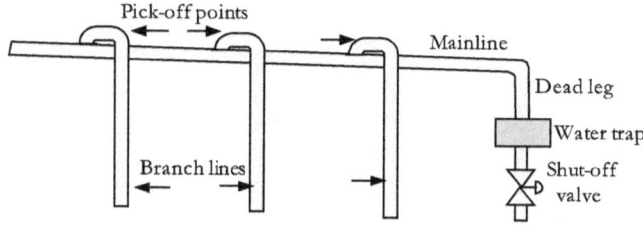

Figure 5.3 | Tapping of branch lines

The distribution pipe system can be considered as a part of the storage, and the compressed air inside is also subjected to external cooling. This cooling causes the moisture in the air to condense and consequently to precipitate water. Hence, to provide drainage, the pipes should be inclined 1 to 2% downward, in the direction of the airflow, preferably to each corner.

Additionally, all take-off points are tapped from the top of the pipe, as shown in Figure 5.3, to prevent the entry of water in the branch lines. The condensate can then be released from the system through a dead leg at the lowest point.

An automatic drain valve can be provided for terminating a dead leg. Accumulated water can then be automatically drained off when pressure is on as well as when the system is shut down.

Heavy demands for compressed air are to be met occasionally at the ends of long lines, which can result in severe pressure loss.

This pressure loss can be avoided by the installation of intermediate reservoirs as close as possible to the demand points.

Typical Tube Specifications
Table 5.1 and Table 5.2 give typical specifications for nylon tubing and polyurethane tubing, respectively.

Table 5.1 | Specifications of nylon tubing

OD (mm)	ID (mm)	Maximum pressure (bar)	Minimum bend radius (mm)
4	2.5	28	25
5	3	31	25
6	4	25	30
8	6	19	50
10	7.5	24	60
12	9	18	75
14	11	15	80
16	12	18	95
22	17	15	125
28	22	15	160

Table 5.2 | Specifications of polyurethane tubing

OD (mm)	ID (mm)	Maximum pressure (bar)	Minimum bend radius (mm)
4	2.5	10	7
5	3	11	9
6	4	9	16
8	5.5	9	17
10	7	9	25
12	8	9	--

Sizing of pipe systems

An under-sized pipe in a pneumatic system produces a significant pressure drop and consequent energy losses. The pipe size should be selected appropriately to keep the pressure reasonably constant over the whole system. An over-sized pipe costs more. The correct sizing of each part of the pipe system should be ensured for reliable, efficient, and trouble-free operation of the air distribution system.

The selection of pipe size is governed by the following factors: (1) Delivery volume of the compressor, (2) Total pipe length, (3) Operating pressure, (4) Permissible pressure drop (maximum 1 bar), (5) Use of fittings such as elbows, T-pieces, and valves.

Pipe manufacturers provide tables or nomograms linking delivery volumes, pipe lengths, operating pressures and permissible pressure drops to different pipe diameters. Selection is best made with the support of nomograms.

An example of finding the pipe diameter using nomograms is illustrated in Example 5.1.

Bends, Couplings, and other Restrictions

Bends, couplings, and other restrictions also increase the pressure drops. The pressure drops in pipe fittings are generally specified in terms of equivalent lengths of a standard pipe. The equivalent pipe lengths are given in Table 5.3.

Table 5.3 | Equivalent pipe lengths in metres

Fitting	Pipe diameter (mm)						
	25	40	50	80	100	125	150
Bend 90°(R=d)	0.3	0.5	0.6	1.0	1.5	2.0	2.5
Bend 90° (R=2d)	0.1	0.2	0.3	0.5	0.8	1.0	1.5
T-piece	2	3	4	7	10	15	20
Check valve	8	10	15	25	30	50	60

Example 5.1

The air consumption in an industrial plant is found to be 240 Nm³/hour. Next, the likely increase in air consumption for about three years is three times the current air consumption. The total length of the pipe is measured to be 280 metres. Additionally, the distribution network contains T-pieces (6 Nos.), normal elbows (5 Nos.) and two-way valve (1 No.). The permissible pressure drop is limited to 0.1 bar, and operating pressure is 8 bar. Calculate the pipe diameter using the given nomograms (Figure 5.4 and 5.5).

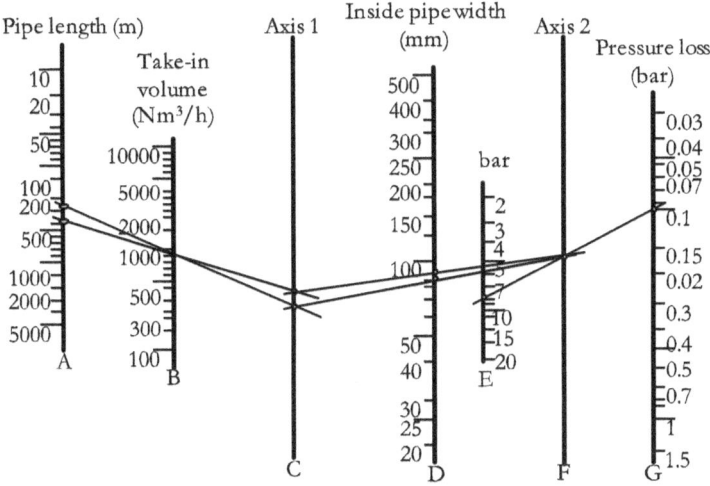

Figure 5.4 | Nomogram to determine the internal pipe diameter
(Not to scale)

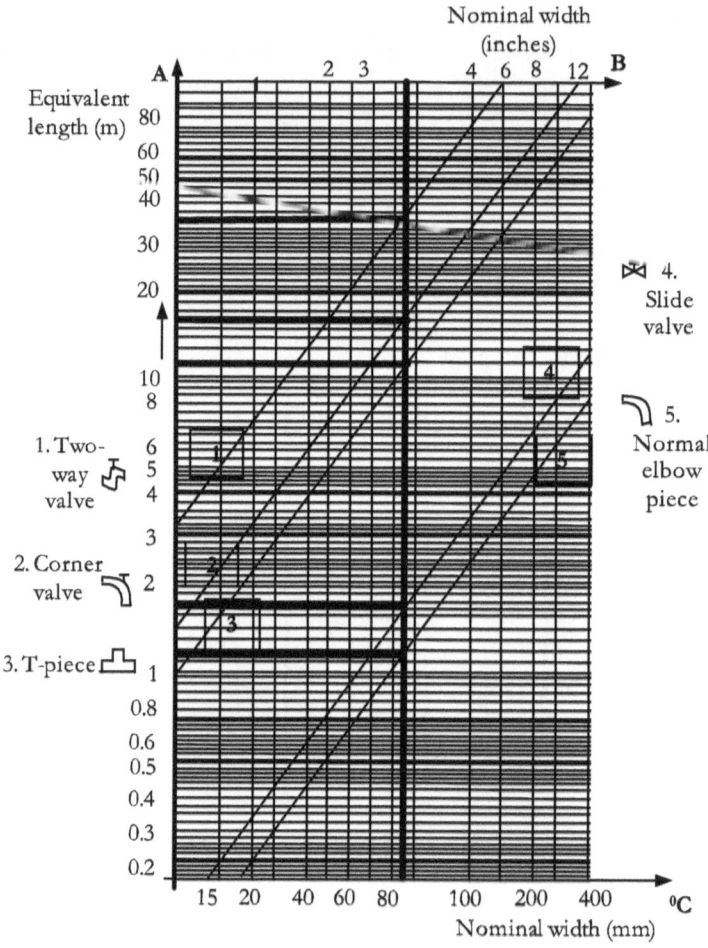

Figure 5.5 | Nomogram for equivalent lengths
(Not to scale)

Given

Total air consumption	= (240 + 720) Nm³/hour
	= 960 Nm³/hour
Length of pipe	= 280 m
Pressure	= 8 bar
Pressure drop	= 0.1 bar

In Figure 5.4, mark and join points corresponding to the pipe length and delivery volume on lines A and B respectively of the nomogram and extend it to line C (Axis 1). Mark and join points corresponding to the pressure and permissible pressure drop on lines E and G, respectively. This line passes through a point on line F (Axis 2).

These points on Axis 1 and Axis 2 are joined together to obtain an intersection at line D. This corresponds to the initial value for the internal pipe diameter. (Say 90 mm)

The resistances are specified in equivalent lengths using the nomogram for equivalent length to take into account the excess pressure drops in pipefitting. Equivalent length is understood to mean the length of a straight pipe having the same resistance to flow as the restrictive element. Equivalent lengths of pipe fittings, corresponding to the initial estimate of pipe diameter (90 mm), can be obtained from nomogram in Figure 5.5 for equivalent length and the obtained values are given below:

6 T-pieces	= 6 x 10 m = 60 m
1 two-way valve	= 1 x 32 m = 32 m
5 elbow pieces	= 5 x 1m = 5 m
Total for fittings and valves = 97 m	
Total pipe length	= (280+97) m = 377 mm

With this modified pipe length, find the pipe diameter again using the nomogram for pipe diameter. In this example, the revised pipe diameter is found to be 95 mm. The same size or the next higher size available in the market can be selected as the pipe diameter.

Chapter 6 | Components for the Secondary Air Treatment

The secondary air treatment is an effort to prepare compressed air finely, regulate pressure to the requirement of the application, and perhaps to mix the air with a fine mist of lubricating oil, just before the entry of compressed air into the associated application. The components used in the secondary air treatment are filter, regulator, and lubricator (FRL).

Filter
A filter can be fitted to the branch line of a compressed air system to remove dust, dirt, and water from the compressed air.

The flow rate requirement of a filter in the branch line is much lower than that of the filter in the mainline.

Pressure regulator
Pneumatic machines and appliances require relatively steady pressure for their satisfactory operation. However, the operating pressure tends to fluctuate due to variations in the supply pressure or load pressure. It is therefore essential to regulate the operating pressure to match the requirements of the load, regardless of the variations in the supply pressure or the load pressure. Diaphragm regulators are the commonly used pressure regulators in industrial pneumatic systems.

Filter-regulator
In this design, filter and regulator are combined as a single unit. Air flows first through the filter and is then directed to the regulator. The advantage of this design is that only one unit is to be mounted, thus simplifying the installation work and reducing the cost.

Pressure gauge
The most commonly used device for measuring pressure is the bourdon tube.

Lubricator

The requirement of lubrication for the moving parts of components in a pneumatic system can be met by using valves and cylinders provided with an integral lubricant in each component or by injecting a controlled quantity of oil mist into the air stream using a mist lubricator.

Lubricators can be of the micro-fog type or the oil-fog type. The micro-fog type is used for the most general-purpose applications, and the oil-fog type is used for heavy-duty applications.

The lubricating oil is usually stored in a transparent polycarbonate bowl or a metal tank of large volume with a sight glass. A micro-fog lubricator cannot be filled under pressure except when fitted with a quick fill device.

Air service unit

An FRL unit is shown in Figure 6.1. It comprises the following:
- Shut off valves to isolate upstream air and downstream air
- A combined filter and pressure regulator unit with a gauge
- Lubricator

For ease of use and system flexibility, a handy method of combining these units is to use a modular system. Typical specifications of pressure regulators and lubricators are given in Table 6.1 and Table 6.2.

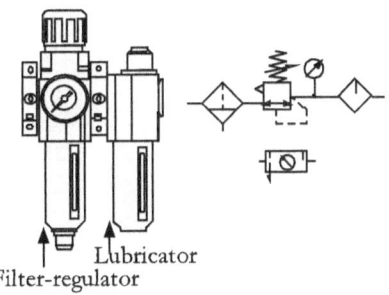

Filter-regulator   Lubricator

Figure 6.1 | Air service unit

Typical Specifications of Pressure Regulators

Table 6.1 | Specifications of pressure regulators

Flow l/s	Operating pressure range (bar)	Size
35	0.3 to 10	G¼
80	0.3 to 10	G⅜
120	0.3 to 10	G½
120	0.3 to 10	G¾
150	0.4 to 8	G¾
180	0.4 to 8	G1
180	0.4 to 8	G1¼
180	0.4 to 8	G1½

Regulator - with gauge / without gauge
Diaphragm: Relieving / Non-relieving
Threads: PTF / ISO Rc taper / ISO G parallel

Typical Specifications of Lubricators

Table 6.2 | Specifications of lubricators

Flow l/s	Bowl capacity (litre)	Operating pressure, bar	Size
25	0.2	0 – 16 bar	G¼
62	0.2	0 – 16 bar	G⅜
72	0.2	0 – 16 bar	G½
72	0.2	0 – 16 bar	G¾
105	0.5	0 – 16 bar	G¾
140	0.5	0 – 16 bar	G1
140	0.5	0 – 16 bar	G1¼
140	0.5	0 – 16 bar	G1½

Type: Oil-fog / Micro-fog
Bowl: Metal with level indicator / Transparent with guard
Threads: PTF / ISO Rc taper / ISO G parallel

Chapter 7 | Pneumatic Actuators

There are two basic types of pneumatic actuators. They are: (1) Linear actuators and (2) Rotary actuators.

Linear Actuators
A pneumatic cylinder converts pneumatic power into a controllable linear force or motion or both.

Terms and Definitions
Some essential parameters concerned with the operation and applications of pneumatic cylinders are its bore diameter, piston-rod diameter, force (thrust and pull), stroke length, speed, and piston-rod buckling.

Maximum operating pressure (P): It is the pressure that overcomes all resistances in the system, which includes both useful work and losses. Alternatively, it is the maximum working pressure that the cylinder can sustain without adverse consequences.

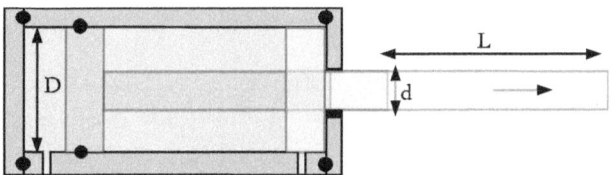

Figure 7.1 | Cylinder parameters

Bore Diameter (D): It refers to the diameter at the bore of the cylinder (Figure 7.1). It can be used to calculate the bore area of the cylinder. It is also equal to the piston diameter, in a close-fitting pneumatic cylinder.

Piston-rod Diameter (d): It refers to the diameter of the piston-rod of the cylinder (Figure 7.1).

Maximum Stroke Length: It is the maximum linear movement that a cylinder can produce. The maximum stroke length for single-acting cylinders is typically 100 mm. The maximum stroke length for double-acting cylinders is 2000 mm. For special designs, the stroke lengths can be up to 10 m.

Thrust

The theoretical thrust (out-stroke) or pull (in-stroke) of a cylinder is the product of the effective piston area and the operating pressure.

The effective area considered for the calculation of thrust is the full area of the cylinder bore and is given by $\pi D^2/4$. The effective area for the calculation of pull is the full area of the cylinder bore minus the rod area and is given by $\pi (D^2-d^2)/4$.

$$\text{Thrust, F (Newton)} = \text{P (Pascal) x } A_{ext} \text{ (m}^2)$$
$$\text{Pull, F (Newton)} = \text{P (Pascal) x } A_{ret} \text{ (m}^2)$$

Where,

A_p is the piston area
A_r is the piston-rod area
A_{ext} is the active area during extension: $(A_{ext} = A_p)$
A_{ret} is the active area during retraction: $(A_{ret} = A_p - A_r)$

The present-day-practice is to specify bore (D) and piston diameter (d) in millimetres and working pressure (P) in bar. In the formula for force, P is divided by 10 to express pressure in Newton per square millimetre [1 bar = $(1/10)$ N/mm^2]. The theoretical force (F) is given by:

$$\text{Thrust, F} = \frac{\Pi D^2}{4} \cdot \frac{P}{10} \text{ Newton}$$

$$\text{Pull, F} = \frac{\Pi (D^2-d^2)}{4} \cdot \frac{P}{10} \text{ Newton}$$

Example 7.1 | A pneumatic double-acting pneumatic press cylinder with an effective piston area of 19.625 cm² for push stroke, and a piston-rod area of 3.14 cm², operating at 6 bar does produce what theoretical forces for the push stroke and pull stroke?

Solution

Piston area, push stroke, A_{push}	$= 19.625$ cm²
Piston Rod area, A_{rod}	$= 3.14$ cm²
Pressure, P	$= 6$ bar

Effective piston area, pull stroke, $A_{pull} = A_{push} - A_{rod}$
$$= (19.625 - 3.14) \text{ cm}^2$$
$$= 16.485 \text{ cm}^2$$

Thrust, F_{push} $\quad = P \times A_{push}$
$\quad\quad\quad\quad\quad\quad = (6 \times 10^5) \times (19.625 \times 10^{-4}) \text{ N}$
$\quad\quad\quad\quad\quad\quad = 1178 \text{ N}$

Pull, F_{pull} $\quad\quad = P \times A_{pull}$
$\quad\quad\quad\quad\quad\quad = (6 \times 10^5) \times (16.485 \times 10^{-4}) \text{ N}$
$\quad\quad\quad\quad\quad\quad = 989 \text{ N}$

Tables 7.1 and 7.2 give the forces of single-acting cylinders and double-acting cylinders, respectively. Figures given in the tables do not make allowance for the loss due to friction or air leakage.

Since the air pressure in a plant may vary erratically, due to intermittent use of large volumes of compressed air in all types of pneumatic equipment, the bore size of the cylinder must be large enough to provide the force required after allowing for any normal pressure drop.

Thrusts and Pulls of Single-acting Cylinders

Table 7.1 | Thrusts and pulls of single-acting cylinders

Cylinder bore, mm	The minimum pull of spring, N	Thrust, N at 6 bar
10	3	37
12	4	59
16	7	105
20	14	165
25	23	258
32	27	438
40	39	699
50	48	1102
63	67	1760
80	86	2892
100	99	4583

Thrusts and Pulls of Double-acting Cylinders

Table 7.2 | Thrusts and pulls of double-acting cylinders

Cylinder bore mm	Piston rod dia (mm)	Thrust, N (at 6 bar)	Pull, N (at 6 bar)
8	3	30	25
10	4	47	39
12	6	67	50
16	6	120	103
20	8	188	158
25	10	294	246
32	12	482	414
40	16	753	633
50	20	1178	989
63	20	1870	1681
80	25	3015	2721
100	25	4712	4418
125	32	7363	6881
160	40	12063	11309
200	40	18849	18095
250	50	29452	28274
320	63	48254	46384

Note: For pressures other than 6 bar, multiply the thrust/pull at 6 bar by the given absolute pressure and divide it by 7.

These figures do not account for the seal or packing friction in these cylinders. This type of friction is estimated to affect the thrust of the cylinders by about 10%.

Cylinder air consumption

The equations for the volume of free air displaced by the piston during the out-stroke and in-stroke of a double-acting cylinder are given below:

$$V(\text{out-stroke}) = \frac{\pi D^2}{4} \ S \ \frac{P_s + P_a}{P_a} \ 10^{-6}$$

$$V(\text{in-stroke}) = \frac{\pi(D^2 - d^2)}{4} \ S \ \frac{P_s + P_a}{P_a} \ 10^{-6}$$

Where,

      D = Cylinder bore, mm
      d = Rod diameter, mm
      V = Volume of free air, dm³
      S = Stroke, mm
      P_s = Supply gauge pressure, bar
      P_a = Atmospheric pressure (assumed to be 1 bar)
      $(P_s + P_a)/P_a$ = Compression ratio

The Compression ratio $(P_s + P_a)/P_a$ may be considered as a multiplying factor in order to normalise the pressure condition.

Example 7.2

Calculate the air consumption per mm stroke of a double-acting cylinder with 32 mm bore and 12 mm piston-rod diameter supplied compressed air at a pressure of 6 bar.

Solution

Bore diameter = 32 mm
Rod diameter = 12 mm

$$V(\text{out-stroke}) = \frac{\pi \cdot 32^2}{4} \ . \ 1 \ . \ \frac{6+1}{1} \ 10^{-6} = 0.00563 \ \text{dm}^3/\text{mm}$$

$$V(\text{in-stroke}) = \frac{\pi \cdot (32^2 - 12^2)}{4} \ . \ 1 \ . \ \frac{6+1}{1} \ 10^{-6} = 0.00484 \ \text{dm}^3/\text{mm}$$

Volume (Total) = 0.01047 dm³ per mm/cycle

Air consumption

Table 7.3 gives the air consumption chart for pneumatic cylinders.

Table 7.3 | Air consumption of pneumatic cylinders

Bore	Rod	Air consumption for the		
		forward stroke of 1 mm at 6 bar	return stroke of 1 mm at 6 bar	combined strokes of 1 mm at 6 bar
mm	mm	dm³/mm	dm³/mm	dm³/mm
10	4	0.00054	0.00046	0.00100
12	6	0.00079	0.00065	0.00144
16	6	0.00141	0.00121	0.00262
20	8	0.00220	0.00185	0.00405
25	10	0.00344	0.00289	0.00633
32	12	0.00563	0.00484	0.01047
40	16	0.00880	0.00739	0.01619
50	20	0.01374	0.01155	0.02529
63	20	0.02182	0.01962	0.04144
80	25	0.03519	0.03175	0.06694
100	25	0,05498	0.05154	0.10652
125	32	0.08590	0.08027	0.16617
160	40	0.14074	0.13195	0.27269
200	40	0.21991	0.21112	0.43103
250	50	0.34361	0.32987	0.67348

1 cubic decimeter (dm³) = 1 litre

Note:
1. *Take each figure and multiply by the stroke in mm.*
2. *For pressures other than 6 bar, multiply the air consumption value by the given absolute pressure and divide it by 7.*

To estimate the total average air consumption of a typical pneumatic system, calculate the air consumption for each cylinder in the system using the formulae given above. Add the estimated air consumption of all cylinders and add 5% to make allowance for the loss due to leakage and friction.

Cylinder Speed

Assume that the piston-rod assembly of a cylinder moves with a velocity of 'v' when pushed by the system fluid with a flow rate 'Q'. Further, assume that the cylinder piston of area 'A' has moved a distance 'S' in time 't' for attaining the velocity v.

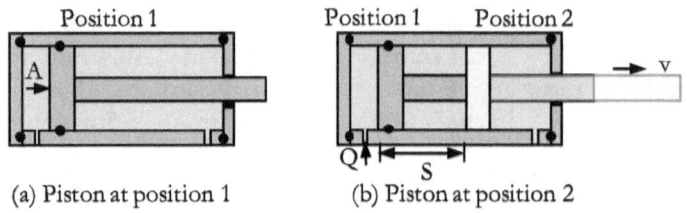

(a) Piston at position 1   (b) Piston at position 2

Figure 7.2 | Illustration of a cylinder in two piston positions

Figure 7.2 (a) and (b) show two positions of a cylinder with the piston in position 1 and position 2 respectively for determining the speed during its forward stroke. Figure 7.2(b) shows the positions '1' superimposed. Mathematically,

$$v = S/t \quad \text{or} \quad t = S/v$$

We can easily relate the theoretical flow rate (Q) of the fluid to the speed (v) at which the piston-rod moves if we consider the cylinder volume (V) that must be filled with the fluid and the distance (S) through which the piston must travel at the specified speed. The volume (V) of the cylinder is the length of the stroke (S) multiplied by the piston area (A). The flow rate (Q) to achieve the required speed (v) is given below.

$$Q \ (m^3/s) = \frac{V \ (m^3)}{t \ (s)} = \frac{A \ (m^2) \ x \ S \ (m)}{t \ (s)} = A \ (m^2) \ x \ v \ (m/s)$$

It can be observed from the equation mentioned above that the speed (v) of a given cylinder depends on the flow rate (Q) of the fluid.

Cylinder buckling

If a compressive axial load is to be applied to a long piston rod, it must be within the safety limit to prevent rod buckling. Due to buckling stress, the permissible load of the cylinder that has a long stroke length is lesser than that ought to have been provided by the same maximum permissible working pressure and the piston surface area. This load should not go beyond certain maximum values as given by Euler's formula that is related to the stroke and piston rod diameter. Euler's formula for elastic instability is expressed as:

$$F_k = \frac{\pi^2 EJ}{l_k^2 S}$$

Where,

F_k = Permissible buckling force (N)
E = Modulus of elasticity (N/mm^2)
J = Moment of inertia (m^4)
S = Safety factor (Chosen as 5)
l_k = Equivalent free buckling length (cm)
= (1 to 2) x stroke length, l

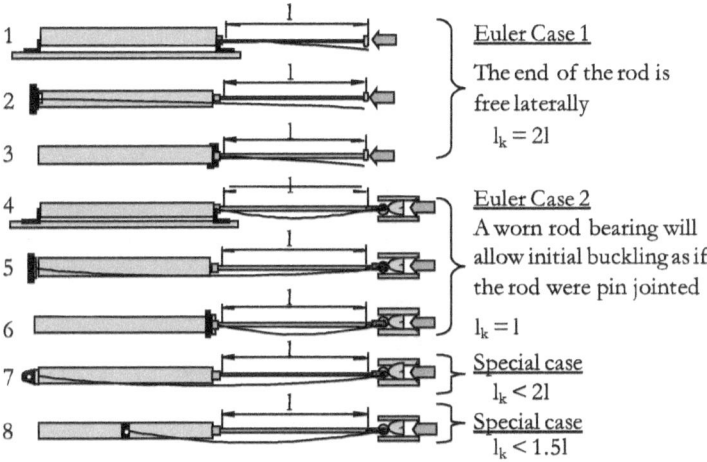

Figure 7.3: Equivalent free buckling length.

The equivalent free buckling lengths for different cylinder mounting arrangements are given in figure 7.3. For a slender column with one end free and other end fixed (Euler case 1), $l_k = 2l$. For a pin-jointed slender column (Euler case 2), $l_k = l$. In certain exceptional cases, $l_k < 2l$ and $l_k < 1.5l$.

Limitations on Maximum Thrust Force
Its piston-rod diameter and overall length limit the maximum thrust force which a cylinder can practically provide. The cylinder must also be supported adequately.

Note that a head-end mounting provides greater column strength than the cap-end mounting, due to the smaller distance between the mounting points in the head-end mounting than that in the cap-end-mounting.

The piston-rod size of a pneumatic cylinder can be selected from the size charts, with the help of values of its free buckling length and the load imposed on the cylinder.

Technical data for pneumatic Cylinders

Standard values of cylinder diameters, maximum stroke lengths, are presented in Table 7.4.

Table 7.4 | Standard values

Parameters	Standard values	Remarks
Bore diameter (mm)	10, 12, 16, 20, 25.	ISO 6432
	32, 40, 50, 63, 80, 100, 125, 160, 200, 250, 320.	ISO 6431
Rod diameter (mm)	1 to 63 mm	
Stroke (mm)	10, 25, 40, 50, 80, 100, 125, 160, 200, 250, 320, 400, 500 ...2000.	-
Operating pressure (bar)	Up to 12 bar	

Materials Used for Pneumatic Cylinders

Table 7.5 gives the materials used for the construction of pneumatic cylinders.

Table 7.5 | Materials used for cylinders

Material	Trade name	Part
Brass	-	Barrel
Steel	-	Barrel
X 5 Cr Ni 18 9	-	Barrel
X 20 Cr 13	-	Piston rods
Cast iron	-	Mounting parts
Acrylonitrile butadiene rubber (NBR)	Perbunan	Seals
Fluroelastomeric rubber (FKM)	Viton	Heat resistant seals
Polyurethane (PUR)	-	Diaphragm

Standards for Pneumatic Cylinders

Table 7.6 gives the standards for pneumatic cylinders.

Table 7.6 | Pneumatic cylinder standards

Pneumatic component	Standard	Remarks
	ISO 6431	International
	ISO 6432	International
Standard cylinder	VDMA 24652	German
	NFE 49003.1	France
	UNI 10290	Italy
	NFPA (JIC)	USA

Rotary actuators

A rotary actuator converts the energy of the compressed air into rotary mechanical energy. Air motors are designed for continuous rotation. Semi-rotary actuators are designed for reciprocating, rotary motion.

A rotary actuator can be defined in terms of the torque it produces and its running speed. The starting torque of a rotary actuator is the torque available to move a connected load from rest. Stall torque is the torque that must be applied by the load to bring a running actuator to rest. Running torque is the torque available at any given speed.

An air motor can withstand repeated stalling and reversing without harm or overheating. It can accelerate rapidly since the energy contained in the compressed air is released at a high rate.

Semi-rotary actuators

Semi-rotary actuators are constructed with a rotating vane or with a rack-and-pinion design. In the vane type rotary actuator as shown in Figure 7.4, with limited travel consists of a single vane coupled to the output shaft. It is usually designed for a double-acting operation with a maximum angle of rotation of 270°. Usually, the angle of rotation can be adjusted.

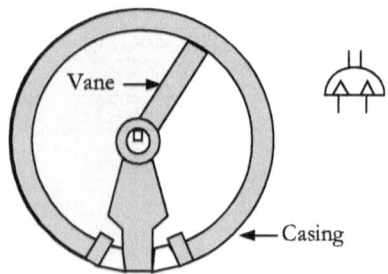

Figure 7.4 | Semi-rotary actuator - Vane type

The rack-and-pinion type of rotary actuator with limited travel is shown in Figure 7.5. It consists of a double-acting piston coupled to the output shaft by a rack-and-pinion arrangement. The angle of rotation up to 360° is possible with this type of design.

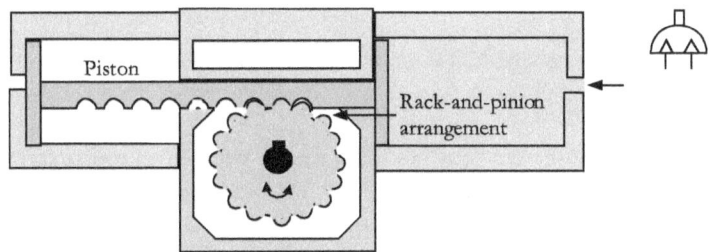

Figure 7.5 | Semi-rotary actuator - Rack-and-pinion type

Air motors
Air motors convert the potential energy of compressed air into rotary mechanical energy. They are designed to provide continuous rotation. Piston, vane and gear designs are generally used for air motors.

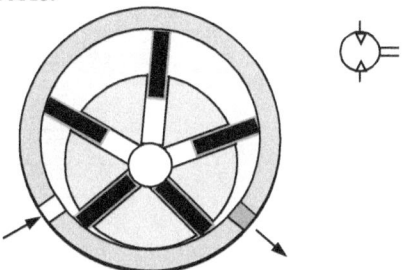

Figure 7.6 | Sliding vane motors

A rotary vane motor is shown in Figure 7.6. It consists of a cylindrical rotor with sliding vanes placed eccentrically in a cylindrical housing. As air enters the inlet port and passes into the cylinder, a pressure unbalance acts on the vanes. This pressure unbalance develops a torque that turns the rotor against the motor's load.

Piston motors have four to six cylinders, which are arranged, in either radial or axial positions. The cylinder pistons reciprocate in sequence when compressed air is applied. This piston movements cause a crankshaft to turn through a connecting rod, thus causing the rotation of the output shaft.

In the gear design, torque is generated by the teeth profiles of two meshed gear wheels.

Typically, air motors are available in the power ratings in the range from $\frac{1}{8}$ to 25 hp. Many speed ranges are possible extending from as low as 40 rpm up to as high as 50000 rpm.

Technical data for Rotary Actuators
Standard values of sizes and rotating angles of rotary actuators are presented in Table 7.7.

Table 7.7 | Standard values

Parameters	Standard values	Remarks
Standard size of rotary actuators	6, 8, 10, 12, 16, 25, 32, 40, 100 mm	-
Standard rotating angle for rotary actuators	90°, 180°, 270°, 360°	Fixed
	30° to 270°	Variable
Torque	0.15 to 150 Nm	
Operating pressure	Up to 12 bar	

Note: Additional data for different types of actuators are given in Appendix 2

Pneumatic tools

Compressed air permits the use of tools, which are compact, light in weight, portable and easy to operate. Pneumatic portable tools carry out a wide range of operations such as nut running, screw driving, grinding, drilling, riveting, scaling, stud driving, and wire wrapping. Portable tools include screwdrivers, hammers, riveters, abrasive tools and hoists. Table 7.8 gives the air consumption chart for industrial type tools.

Air Consumption Chart for Industrial Type Tools

Table 7.8 | Air consumption at 5 to 6 bar

Tool	Consumption (lpm) at 25% usage factor
Air motor, 1 hp to 3 hp	255 to 680
Burring tool, small to large	113 to 170
Chipping hammer	227
Die grinder, medium	170
Drill, 1/16" to 5/8"	170 to 255
Horizontal grinder, 2", 4", 6", 8"	227 to 566
Impact Wrench, 1/4" to 1¼"	113 to 396
Nut setters, small up to 3/8" to 3/4"	170 to 425
Paint spray gun	142
Rammers, small, medium, large	170 to 283
Riveting Hammer, Light, heavy	113 to 227
Saws, circular	453
Scaling hammer	85
Screwdriver	85 to 170
Trapper, up to 3/8"	170
Vertical grinders and sanders	255 to 566

- *Air consumption is only indicative and may not be accurate for any particular make*
- *Always check with the OEM for the actual air consumption of tools*
  Note: Additional data for pneumatic tools are given in Appendix 3

Vacuum Grippers

A vacuum gripper consists of a vacuum generator (ejector) and a suction cup. An air suction or vacuum filter is provided in the vacuum passage for preventing the intrusion of dust into the ejector. Figure 7.7 shows a vacuum generator and suction cups.

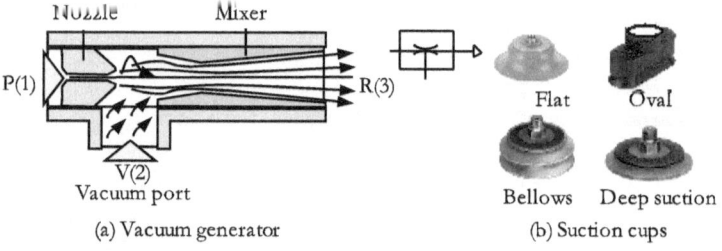

(a) Vacuum generator   (b) Suction cups

Figure 7.7 | Vacuum generator and Suction cups

Vacuum Gripper Terms

Maximum Suction Flow Rate: It is the maximum value of the volume of air taken in without having anything connected to the vacuum port.

Maximum vacuum pressure: It is the maximum value of the vacuum pressure generated by the ejector.

Air consumption: It is the volume of the compressed air consumed by the ejector.

Typical Specifications, Vacuum Generators

Table 7.9 gives typical specifications of vacuum generators.

Table 7.9 | Typical specifications of vacuum generators

Nozzle diameter (mm)	Max. suction flow (Nlpm)	Air consumption (Nlpm)
0.5	6	13
1.0	26	52
1.3	40	84
1.5	58	113
1.8	76	162
2.0	90	196

Suction Cups

Suction cups are used for the transportation of work-pieces of different weights, surfaces, and shapes.

They are made from nitrile rubber, polyurethane, and silicone for use in a wide variety of applications. Suction cups made from silicone are food-safe.

Round flat type suction cups are capable of retaining work-pieces with smooth, impervious surfaces. However, round bellows-type suction cups can adapt to suit uneven, curved, and inclined surfaces.

Typical nominal diameters for suction cups are as follows: 2, 5, 8, 10, 15, 30, 40, 55, 75, 100, and 125 mm.

Typical Specifications of Round Flat Suction Cups

Table 7.10 gives typical specifications of suction cups.

Table 7.10 | Typical specifications of round flat suction cups

Suction cup diameter, mm	Holding force, (N)	Vacuum connection
2	0.14	M3
5	0.9	M5
8	1.6	M5
10	4.5	M5
15	7.9	G⅛
30	34	G⅛
40	56	G¼
55	106	G¼
75	197	G¼
100	397	G¼
125	606	G⅜

Holding force at nominal operating pressure -0.7 bar

Chapter 8 | Pneumatic Valves

A pneumatic valve consists of a body with an internal moving element, such as a poppet or spool, actuating mechanisms, and many ports. It is a control device that directs or restricts the compressed air flow, or controls the flow based on a specified pressure condition in a particular part of the associated circuit. Accordingly, pneumatic valves can be classified as directional control valves, flow control valves, and pressure control valves.

Directional Control Valves
A directional Control (DC) valve (or way valve) controls the path taken by the compressed air. A non-return valve (NRV) is a kind of directional control valve which allows the flow of the compressed air in a pneumatic system in only one direction and blocks the flow in the opposite direction.

3/2-DC Valves (NC Type)
3/2-way normally-closed type valves are used as final control elements to control single-acting cylinders, unidirectional motors, and other valves. The symbol of a 3/2 DC valve is given in Figure 8.1.

5/2-Directional Control Valves
5/2-DC valves are used as final control elements to control double-acting cylinders. The symbol of a 5/2 DC valve is given in Figure 8.1.

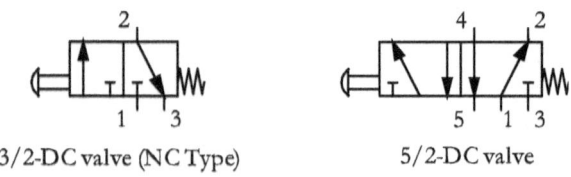

3/2-DC valve (NC Type)   5/2-DC valve

Figure 8.1 | Symbols of directional control valves

Check Valves and Flow Control Valves

A flow control valve restricts the flow rate at which pressurised fluid is transferred in a pneumatic circuit. The symbols of a check valve and a one-way flow control valve are given in Figure 8.2.

Check valve   One-way flow control valve

Figure 8.2 | Symbols of a check valve and a flow control valve

Pressure Control Valve

A pressure sequence valve generates a control signal when a set pressure in a particular part of the system has been reached for initiating a subsequent action.

Pneumatic Valve Sizing

There are many methods that aid in the selection of pneumatic valves used as final control elements. Two prominent methods of sizing pneumatic valves are: (1) finding the value of the coefficient of flow Cv (or Kv) parameter or (2) using sizing table that relates Cv values to the cylinder bore sizes.

The value of Cv (valve coefficient) required to operate a given cylinder for a specific time is given below.

$$C_v = \frac{\text{Area x Stroke x A x } C_f}{\text{Time x 29}}$$

Where,
Cylinder area in square inch
Cylinder stroke in inch
A = Pressure drop constant (From Table 8.1)
C_f = Compression factor (From Table 8.1)
Time in seconds

The compression factor C_f and the pressure drop constant A can be found from Table 8.1.

Table 8.1 | Compression factor C_f and Pressure drop constant A

Inlet pressure psi	Compression factor, C_f	Constant A at the pressure drop		
		2 psi	5 psi	10 psi
10	1.6	--	0.102	--
20	2.3	0.129	0.083	0.066
30	3.0	0.113	0.072	0.055
40	3.7	0.097	0.064	0.048
50	4.4	0.091	0.059	0.043
60	5.1	0.084	0.054	0.040
70	5.7	0.079	0.050	0.037
80	6.4	0.075	0.048	0.035
90	7.1	0.071	0.045	0.033
100	7.8	0.068	0.043	0.031
110	8.5	0.065	0.041	0.030
120	9.2	0.062	0.039	0.29

For most applications, the constant A is selected for a pressure drop of 5 psi. The higher the Cv, the greater the flow. To account for losses, oversize a valve by at least 25%.

The sizing chart in Table 8.2 indexes the Cv values to the cylinder bore sizes and the operating speeds (inches/s) of the cylinder.

Table 8.2 | Sizing chart for pneumatic valves

Cv	Bore size									
	0.75	1.13	1.5	2.0	2.5	3,25	4.0	5.0	6.0	8.0
0.1	26.8	11.9	6.7	3.8	2.4	1.4	0.94	0.6	0.42	0.24
0.2	53.7	23.9	13.4	7.5	4.8	2.9	1.9	1.2	0.84	0.47
0.5	134	59.6	33.6	18.9	12.1	7.1	4.7	3	2.1	1.2
1.0	268	119	67.1	37.7	24.2	14.3	9.4	6	4.2	2.4
2.0	537	239	134	75.5	48.3	28.6	18.9	12.1	8.4	4.7
4.0		477	268	151	96.6	57.2	37.7	24.2	16.8	9.4
8.0			536	302	193	114	75.5	48.3	33.6	18.9
16				604	387	229	151	96.6	67.1	37.7
32					773	457	302	193	134	75.5

Assumption: P = 80 psi, ΔP = 80%

The selection of a pneumatic valve is dependent on the speed requirement of the associated cylinder. A pneumatic valve is designed with the nominal flow rate as an essential parameter. This parameter must be sufficient to meet the air consumption requirement of the cylinder for its satisfactory operation. It may be noted that the port size of the valve should match the port size of the associated cylinder. Another critical parameter for the valve selection is the permissible pressure drop across the valve.

The empirical values for the nominal flow rate provided by the manufacturers can be used as a guide for dimensioning pneumatic valves. Typical values of nominal flow rates are given in Table 8.3. By using these figures, it is possible to achieve cylinder speeds, which are sufficient in most practical cases.

Table 8.3 | Standard nominal flow rate for valves

Cylinder piston dia (mm)	Valve connection size	Approx. Nominal size (mm)	Approx. Standard nominal flow rate (Nlpm)
Up to 25	M5	2.5	105
25 – 50	G⅛	3.5	Up to 180
50 – 100	G¼	7.0	Up to 1140
100 – 200	G½	12.0	Up to 3000
200 – 320	G¾/G1	18.0	Up to 6000

Its desired function mainly decides the selection of other pneumatic valves used in signal processing in a control circuit. For most applications using trip cam and pilot valves, G⅛ poppet models are suitable. Where space is at a premium and the valve being controlled is relatively small, M5 models will suffice.

Chapter 9 | Steps for Pneumatic System Design

A pneumatic system includes many critical components, such as a compressor, actuators, and control valves, and these components are interconnected using pipes and tubes. The energy transfer takes place through the medium of compressed air.

The design of a pneumatic system involves the determination of the force and speed requirements of all actuators, the selection and sizing of components, the air consumption rate of actuators, finding the sequence of operation, the determination of the required pressure level, and finding many other component parameters to meet the design objectives.

The optimum design of a pneumatic system for a project must try to synchronise with the availability and quality of components in the market.

Critical Design Steps

Building the right pneumatic system for the specific application requirements is best achieved by first determining the parameters of the components of the system. That is, the design process of a pneumatic system primarily consists of finding the sizes/capacities of various system components such as cylinders, rotary actuators, valves, compressor(s), drive motor, tank, coolers, filters, dryers, and fluid conductors. The following critical steps may be followed for finding the significant parameters while designing a pneumatic system with a compressor and cylinder.

An analysis of the system to be designed would reveal the application requirements of output force/torque, speed, and mechanical power output. It is also required to identify the nature of each load to be moved in the system and establish the useful load to be handled by each actuator in the system, by applying sound engineering principles. The effects of friction, gravity, mass, inertia, and other external forces, which are likely to be encountered in the system, must be considered. The power transmission mechanism between each load and the corresponding actuator output shaft must be taken into account while calculating the useful load.

Select the operating pressure for a pneumatic system, within the standard pressure range, making allowances for the potential pressure losses and friction in the system. The most economical pressure for industrial pneumatic systems is considered as 6 bar. Most actuators and air tools require 6 to 7 bar. However, many other actuators and tools may require higher pressures up to 16 bar.

Determine the size of the cylinder by taking into account the force, pressure, speed and stroke requirements of the cylinder, and the nature of the load. The dimensions of the cylinder should be such that the force developed by it must overcome the load and frictional forces. The cylinder must also provide adequate acceleration. A single-acting cylinder can be used where the work operation is to be carried out only in one direction of motion of the cylinder. A double-acting cylinder should be used where a positive pneumatic return action is required, and the cycle time is critical. Determine the area and the diameter (bore size) of the cylinder using the formulae given in page 70.

Also, decide the diameter of the cylinder piston-rod. A large piston-rod gives a stronger column strength and compression strength for the piston-rod.

Determine the required stroke length of the cylinder based on the data derived from the system analysis.

Cylinder extension and retraction speeds can be decided as per the system requirements.

Determine the air consumption rate of the cylinder to meet the speed requirements of an application. To estimate the total average air consumption of a pneumatic system, calculate the air consumption for each cylinder in the system using the formulae given in page 74. Values of air consumption for forward and return strokes of pneumatic cylinders are given in Table 7.3 in Page 75. Add the estimated air consumption of all cylinders and add 5% to make allowance for the loss due to leakage and friction.

Determine the air consumption rate of various tools, vacuum generators, and air motors to meet the speed requirements of the application.

Once the air requirement of each of the air consuming devices has been determined, calculate the total air requirement of all the air consuming devices for continuous or intermittent operations.

Find the duty cycle of the application. It is an attempt to find the number of hours the compressor is expected to run. For example, if 70% of the time the compressor is expected to provide air to the equipment, the amount of air to be supplied by the compressor would be 0.7 times the total air requirement as calculated in step 2.

Add an extra 30% to provide a reasonable buffer. Also, consider the requirements for future expansion.

However, it may be noted that the total air consumption, as calculated above, may give rise to excessive capacity requirements in the system. Therefore, one may have to be careful in determining the maximum flow rate expected under the actual service conditions.

Next, select a compressor and drive motor with sufficient delivery rate to meet the peak air consumption rate of all actuators. Draw the important compressor specification parameters such as the drive motor power, drive speed, pressure rating, number of compression stages, operating voltage, etc. Remember, the rule of thumb that on a steady pumping, a compressor will produce a minimum of 113 Nlpm flow of air for every HP capacity at 6 bar.

Select the size of the air receiver tank from the sizes offered by the compressor manufacturer.

Select the specifications of components such as aftercooler, mainline filter, dryer(s), and FRL unit(s), based on the flow rate.

Design the mainline and distribution conductor system to keep the pressure drop across the conductor system to a permissible limit. The selection of pipe size is governed by the delivery volume, required pipe length, operating pressure and the permissible pressure drop.

Next, size all final control elements to meet the air consumption requirement of the actuators, using charts/data provided by manufacturers.

If the velocity of the piston-and-piston-rod assembly of the cylinder is high enough to cause end-of-stroke impact shocks or if there is no speed control feature incorporated for the cylinder, cushioning devices are essential for the cylinder. Typically cushioning is required if the velocity of the cylinder is greater than 0.1 m/s.

The cylinder's permissible piston-rod buckling force is an important parameter to be taken into account while selecting the cylinder.

Next, five sample design problems are presented to highlight the essential design steps. However, it may be remembered that the optimum design depends on the site conditions and exact data available in the manufacturer's domain.

Design Problem 9.1

A service station uses the following types of tools and equipment with the average requirement of compressed air, as given in Table 9.1. Determine the size of the compressor, rating of the motor, and the tank size.

Table 9.1 | Data and calculations for the design problem 9.1

Tool	Qty	Air consumption per tool (lpm)	Total air consumption [lpm(fad)]	Max. pressure (bar)
Blow Gun	1	70.8	70.8	6
Car Lift, 8000 lb	2	169.92	339.84	10
Grease Gun	2	84.96	169.92	8.6
Spark Plug Cleaner	1	141.6	141.6	6
Spring Oiler	1	113.28	113.28	6
Tire Inflation Line	1	56.64	56.64	8.6
Transmission Flusher	1	84.96	84.96	6
		Total	**977.04**	

Solution

Total air consumption = 977.04 lpm (Free air)
Operating pressure = 6 bar
Pressure, lower setting = 7 bar
Pressure, upper setting = 8 bar

Select the compressor specifications from manufacture's chart (See Table A1.3, Page 107)

- Compressor type = Rotary screw
- No. of stages = One
- Power, compressor = 7.5 kW
- Voltage = 440 V, 3 phase
- Maximum pressure = 10 bar
- Tank size = 302.8 litres

Design Problem 9.2

A service station uses the following types of tools and equipment with the continuous requirement of compressed air, as given in Table 9.2. Determine the size of the compressor, rating of the motor, and the tank size.

Table 9.2 | Data and calculations for the design problem 9.2

Tool	Qty	Air consumption per tool (lpm)	Total air consumption [lpm(fad)]	Max. pressure (bar)
Air hammer	1	110	110	7
Body polisher	1	565	565	7
Paint spray gun (Production)	1	225	225	7
Touch-up spray gun	1	100	100	7
Vacuum cleaner	1	200	200	10
		Total	**1200**	

Solution

Total air consumption = 1200 lpm (Free air)
Operating pressure = 7 bar
Pressure, lower setting = 8 bar
Pressure, upper setting = 9 bar

Select the compressor specifications from the manufacture's chart (See Table 3.3, Page 30)

- Compressor type = Reciprocating
- No. of stages = Single-stage
- Power, compressor = 7.5 kW
- Voltage = 440 V, 3 phase
- Maximum pressure = 9 bar
- Tank size = 220 litres

Design Problem 9.3

A double-acting cylinder is used for the clamping operation of a casting during machining. The cylinder has a bore diameter of 63 mm and a stroke length of 200 mm. On average, the cylinder is to clamp 16 castings every minute. The total air consumption of the cylinder for the forward and return strokes can be taken as 0.043103 litre/mm per cycle at 6 bar. The system pressure is set at 7 bar. Find the air consumption of the cylinder at 7 bar.

Solution

Stroke length   = 200 mm

No. of operations/minute = 16

Total air consumption   = 0.043103 litre/mm per cycle at 6 bar

                                  =0.043103 x 200 litre/cycle at 6 bar

                                  = 8.6206 litre/cycle at 6 bar

                                  = 8.6206 x 16 lpm at 6 bar

                                  = 137.9296 lpm at 6 bar

                                  = 137.9296 x (7+1) / (6+1) lpm at 7 bar

                                  = 157.634 lpm of free air at 7 bar

[Refer to pages 74 and 75 for the basics of air consumption of pneumatic cylinder]

Design Problem 9.4

A manufacturing plant has 10 Nos. of automatic production machines using the following types of pneumatic actuators. The operating pressure is set at 6 bar. The duty cycle can be taken as 70%. Determine the size of the compressor, rating of the motor, and the tank size. Assume, 8 operations (average) per minute for each linear actuator.

Table 9.3 | Data for the design problem 9.4

Qty	Air consumption per mm stroke per actuator (l/mm)	Air consumption per cycle per actuator (l/cycle)	Total air consumption for 8 cyclic operations (average) and for all actuators per machine [lpm(fad)]
Single-acting cylinder, 25 mm bore dia, 50mm stroke			
2	0.00344	0.172	0.172x8x2=2.752
Double-acting cylinder, 40 mm bore dia, 50 mm stroke			
3	0.01619	0.8095	0.8095x8x3=19.428
Double-acting cylinder, 50 mm bore dia, 100 mm stroke			
2	0.02529	2.529	2.529x8x2=40.464
Vacuum generator			
4	-	30	120
			182.644

Solution

Air consumption per machine at 100% duty cycle = 182.644 lpm
Actual air consumption for 10 machines = 1826.44 lpm
Actual consumption per machine at 70% duty cycle = 1279 lpm

Select compressor specifications from the chart extracted from manufacturer's domain (Table 3.3, Page 30)

- Compressor type   = Reciprocating piston
- No. of stages   = Single
- Power, compressor   = 11 kW
- Voltage   = 440 V, three-phase
- Maximum pressure   = 9 bar
- Tank size   = 500 litres

Design Problem 9.5

A manufacturing plant requires 10 Nos. of pneumatically-operated production machines to be connected to a ring-main distribution system. Each machine has to be designed for carrying out various work operations with the following specifications:

(1) Work operations, single-acting, 1500 N thrust force, 50 mm stroke, and 4 cycles per minute - 4 operations, and
(2) Work operations, double-acting, 2500 N thrust force, 50 mm stroke, and 6 cycles per minute - 8 operations

The operating pressure is to be set at 6 bar. The duty cycle can be taken as 70%. The compressor has to be located at a distance of 300 m from the ring main. The permissible pressure drop is 0.5 bar. Design a pneumatic system.

Solution		
Operating pressure (P)	6 bar	
No. of machines	10	
For each machine:		
Single-acting cylinders		
No. of single-acting cylinders	4 Nos	
Force, to be developed by each single-acting cylinder (F)	1500 N	
Area of the piston (A = F/P)	2500 mm²	
Bore diameter, D = √(4 * A /∏)	56 mm	
Std bore diameter, (Table 7.2, Pg 73)	63 mm	
Rod Diameter, Selection	20 mm	
Area of the piston, revised	3116 mm²	0.00312 m²
Stroke length (Given)	50 mm	0.05 m²
Air consumption per cycle per cylinder	1090483 Nmm³	0.0011 Nm³
Air consumption per cycle per cylinder (Also refer to Table 7.3, Pg 75)	1.09048275 Nlitre	
No. of cycles of operations per minute	4	
Air consumption per minute per cylinder	4.361931 Nlpm	
Total air consumption of all single-acting cylinders, per machine	17 Nlpm	

Double-acting cylinders		
No. of double-acting cylinders	8 Nos	
Force, to be developed by each single-acting cylinder	2500 N	
Area of the piston (A = F/P)	4167 mm²	
Bore diameter, $D = \sqrt{(4 * A /\prod)}$	73 mm	
Std Bore diameter, (Table 7.2, Pg 73)	80 mm	
Rod diameter, selection	25 mm	
Area of the piston, revised	5024 mm²	0.005024m²
Effective area on the piston-rod side	4533 mm²	
Stroke length (Given)	50 mm	0.05 m²
Air consumption per cycle per cylinder*	3345081 Nmm³	0.003345 Nm³
Air consumption per cycle per cylinder (Also refer to Table 7.3, Pg 75)	3.34508125 Nlitre	
No. of cycles of operations per minute	6	
Air consumption per minute per cylinder	20.07 Nlpm	
Total air consumption of all double-acting cylinders, per machine	161 Nlpm	
Cushion	As required	
Stop tube	As required	
Dual pistons	As required	
Type [ISO / NFPA]	ISO	
Bore section [Square / Round]	Round	
Cylinder mounting [Fange, Foot, Pivot, Clevis, Trunnion	Flange	
Total Air Consumption for all machines		
Total air consumption of all actuators, per machine	178 Nlpm	
Total air consumption of all 10 machines	1780 Nlpm	107 Nm³/hr
Duty cycle	70%	0.7
Effective requirement of air	1246 Nlpm	75 Nm³/hr
Leakage (Assumed)	30%	0.3
Total air consumption of all machines, considering leakage	1620 Nlpm	97 Nm³/hr

*To estimate the total average air consumption of a pneumatic system, calculate the air consumption for each cylinder in the system using the formulae given in page 74. Values of air consumption for forward and return strokes of pneumatic cylinders are given in Table 7.3 in Page 75.

Compressor Unit & Drive Motor		
Working pressure	8 bar	
Operating pressure	6 bar	
Actual Delivery (Nlpm / FAD)	1620 Nlpm	
No. of stages of the compressor	Single-stage	
Type of compressor	Piston type	
The temperature of outlet air	120°C	
Electric Drive, compressor	440 V, 3 Ph	
Drive power (Selection) See Page 30	15.0 kW,	
Drive Speed (Selection)	925 rpm	
Duty cycle	Intermittent	
Drive, coupling	Belt	
Lubrication	Lubricated	
Design	Open frame	
Noise level	70 dBA	
Air receiver Capacity See Page 30	450 litre	
Air receiver type	Horizontal	
Receiver drain	Manual	
Receiver material	M S	
Certification for tank (as per standard)	Yes	
Aftercooler		
Aftercooler required	Yes	
If required, type of configuration	Stand alone	
Type of cooling medium	Air-cooled	
Flow rate capacity@6 bar (at least)	1620 Nlpm	
Approach temperature See Page 44	t_{amb} + 11°C	
Compressor inlet temperature Page 44	120°C	
Maximum temperature	175°C	
Fan rating (Selection)	0.5 hp	
Voltage	230 VAC, 1-ph	
Mainline Filter		
Filter Type	General-purpose	
Filter bowl	Transparent	
Flow rate capacity@6 bar (at least)	1620 Nlpm	27 l/s
Filter element size See Page 47	40 μ	
Service life indicator, filter	Mechanical	
Drain, for filter	Manual	
Working pressure, filter	8 bar	
Connection size, filter	G¼	

Dryer		
Dryer required or not	Required	
Type of dryer	Desiccant type, Twin tower	
ISO Class (Customer requirement)	5	
Flow rate capacity@6 bar (at least)	1620 Nlpm	1.620 Nm^3/min
Pressure dew point	+7°C	
Working Pressure	8 bar	
Inlet air temperature	36°C	
Heater capacity	NA	
Blower capacity	NA	
Connection size, dryer	G¼	
FRL		
Filter		
Filter Type	General-purpose	
Filter bowl	Transparent	
Flow rate capacity@6 bar (at least)	162 Nlpm	2.7 l/s
Filter element size	25 μ	
Service life indicator, for filter	Mechanical	
Drain, for filter	Manual	
Working pressure, filter	8 bar	
Connection size, filter	G¼	
Thread type	ISO G	
Regulator		
Regulator diaphragm type	Relieving type	
Requirement of gauge	With gauge	
Flow rate capacity of the regulator	162 Nlpm	
Pressure rating	8 bar	
Connection size, (Table 6.1, Page 68)	G¼	
Thread type	ISO G	
Lubricator		
Lubricator type	Micro-fog	
Lubricator bowl	Transparent	
Bowl capacity (Table 6.2, Page 68)	0.2 litre	
Flow rate capacity of lubricator	162 Nlpm	
Pressure rating	8 bar	
Connection size	G¼	
Thread type	ISO G	

Final Control Elements for Single-acting Cylinders		
Type	3/2-way Double solenoid valve	
Nominal flow rate, min	162 Nlpm	
Valve connection size	G¼	
Approximate Nominal Size	2.5 mm	
Pressure rating (at least)	8 bar	
Seals	NBR	
Mounting Interface [DIN 24340 A6 / ISO 4401 / CETOP RP 121-H / NFPA D03]	ISO 4401	
Control voltage	24 V DC	

Final Control Elements for Double-acting Cylinders		
Type	5/2-way Double solenoid	
Nominal flow rate, min	162 Nlpm	
Valve connection size	G¼	
Approximate Nominal Size	2.5 mm	
Pressure rating (at least)	8 bar	
Seals	NBR	
Mounting Interface DIN 24340 A6 / ISO 4401 / CETOP RP 121-H / NFPA D03	ISO 4401	
Control voltage	24 V DC	

Conductor (Mainline)		
Total air consumption	1620 Nlpm	97 Nm³/hr
Length of pipe	300 m	
Pressure	8 bar	
Permissible pressure drop	0.5 bar	
Pipe diameter, the initial calculation	38 mm	See Figure 5.4, Page 63
No. of elbows	5 No	
No. of T-pieces	6 No	
No. of 2-way valves	1 No	
No. of corner valves	0	
No. of slide valves	0	
The equivalent length of one elbow	0.25 m	See Figure 5.5, Page 64
The equivalent length of one T-piece	2 m	See Figure 5.5
The equivalent length of one 2-way valve	7 m	See Figure 5.5
The equivalent length of corner valves	0	
The equivalent length of slide valves	0	
The equivalent length of all elbows	1.25 m	x5
The equivalent length of all T-piece	12 m	x6
The equivalent length of all 2-way valve	7 m	x1
The equivalent length of 2-way valves	0	
The equivalent length of 2-way valves	0	
Total of equivalent lengths	20 m	
Total of pipe length + equivalent lengths	320 m	
Pipe diameter, the final calculation	40 mm	See Figure 5.4
Pipe diameter, a selection from manufacturer's domain (Next same or larger size available)	40 mm	

Figure 9.1, Page No. 103, gives the component-level layout of the system showing the essential design parameters.

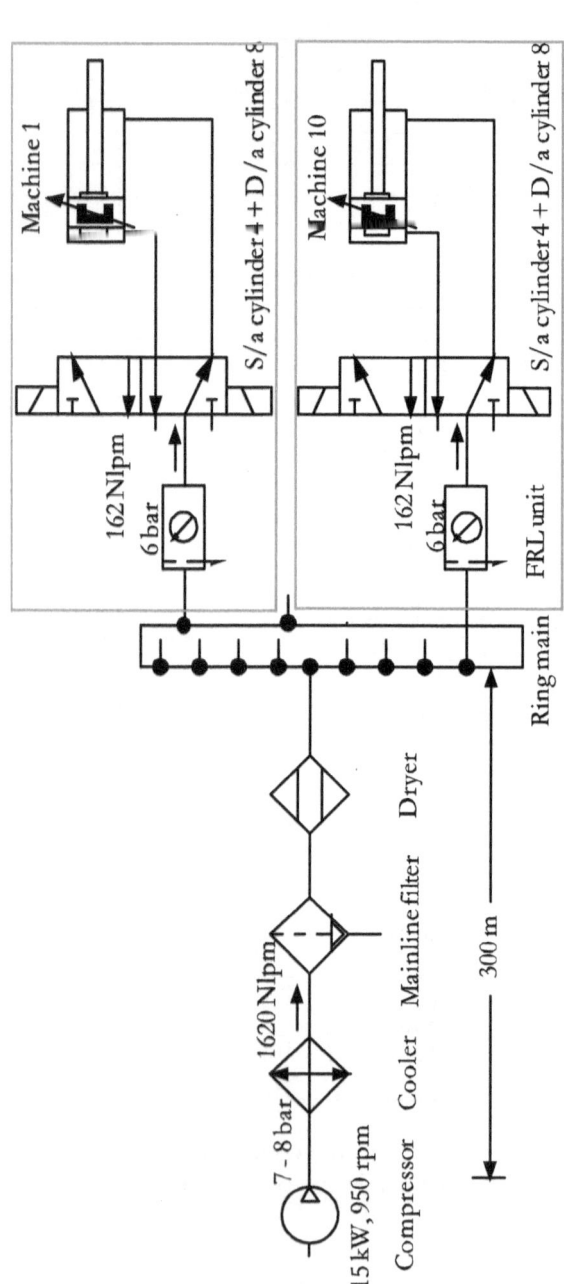

Figure 9.1

Appendix 1

Air Compressor Specifications

Specifications of Portable Electric Air Compressors

Table A1.1 | Specifications of portable compressors

kW	FAD lpm @6 bar	Tank size (litre)	Max. pressure (bar)	Phase (Electric supply)	Input voltage (VAC)	Remarks
0.2	17.0	7.6	7	1	115	--
0.2	21.2	3.8	9	1	115	--
0.4	24.1	7.6	9	1	115	--
0.7	66.6	8.7	9	1	115	--
0.6	73.6	11.4	9	1	115	--
0.6	73.6	22.7	10	1	115	--
0.7	73.6	22.7	10	1	115	--
0.8	85.0	9.5	14	1	115	--
1.5	93.5	9.8	9	1	115	--
0.8	96.3	15.1	9	1	115	--
1.5	107.6	15.9	9	1	115	--
1.1	113.3	5.7	9	1	115	--
0.8	113.3	15.1	9	1	115	--
1.5	116.1	16.3	9	1	115	--
1.5	118.9	12.1	9	1	115	--
1.5	130.3	15.1	14	1	115	--
1.0	130.3	15.1	9	1	115	--
1.3	138.8	22.7	9	1	115	--
1.3	138.8	75.7	9	1	120	H
1.3	138.8	98.4	10	1	120	V
1.5	138.8	113.6	9	1	115/230	H
1.5	138.8	227.1	9	1	115/230	V
1.2	141.6	17.0	14	1	115	--
1.2	141.6	56.8	14	1	120	H
1.2	141.6	56.8	14	1	120	V
1.5	141.6	75.7	14	1	120	V
1.3	144.4	56.8	16	1	120	V
1.5	144.4	90.8	9	1	120	V
1.5	155.8	56.8	9	1	120/240	H
1.5	155.8	75.7	9	1	120/240	H
1.5	155.8	75.7	9	1	120/ 240	V

1.5	155.8	113.6	9	1	120/ 240	V
1.3	155.8	34.1	9	1	120 , 240	H
1.1	178.4	29.5	9	1	120/ 240	H
1.9	184.1	20.1	9	1	115	--
2.2	184.1	19.7	10	1	115	--
1,1	206.7	75.7	9	1	120	H
1.1	206.7	34.1	9	1	120/ 240	H
2.2	288.9	75.7	9	1	240	H
2.2	288.9	227.1	9	1	230	V
2.2	291.7	227.1	9	1	230	V
2.2	359.7	34.1	10	1	240	H
2.2	257.7	113.6	9	3	200-230/ 460	H
3.7	402.1	227.1	10	1	230	V
3.7	439.0	227.1	9	1	230	V
3.7	532.4	34.1	10	1	240	H
3.7	402.1	227.1	10	3	208-230/ 460	V

H – Horizontal | V – Vertical

Specifications of Duplex Air Compressors

Table A1.2 | Specifications of Duplex compressors

kW	FAD lpm @6 bar	Tank size (litre)	Max. pressure (bar)	Phase (Electric supply)	Input voltage (VAC)	
3.7	928.9	454.2	12	1	208-230	H
3.7	928.9	454.2	12	3	208-230/ 460	H
5.6	1387.7	454.2	12	1	208-230	H
5.6	1387.7	454.2	12	3	208-230/ 460	H
7.5	2039.0	757.1	12	3	208-230/ 460	H
11.2	2832.0	757.1	12	3	208-230/ 460	H

H – Horizontal | V – Vertical

Specifications of Rotary Screw Single-stage Compressors

Table A1.3 | Specifications of rotary screw compressors

kW	FAD lpm @6 bar	Tank size (litre)	Max pressure (bar)	Phase (Electric supply)	Input voltage (VAC)	
3.7	515.4	302.8	10	1	230	H/V/E
3.7	515.4	302.8	10	3	208	H/V/E
3.7	515.4	302.8	10	3	230	H/V/E
3.7	515.4	302.8	10	3	460	H/V/E
5.6	764.6	302.8	10	1	230	H/V/E
5.6	764.6	302.8	10	3	208	H/V/E
5.6	764.6	302.8	10	3	230	H/V/E
5.6	764.6	302.8	10	3	460	H/V/E
7.5	1019.5	302.8	10	3	208	H/V/E
7.5	1019.5	302.8	10	3	230	H/V/E
7.5	1019.5	302.8	10	3	460	H/V/E
11.2	1302.7	302.8	10	3	208	H/V/E
11.2	1302.7	302.8	10	3	230	H/V/E
11.2	1302.7	302.8	10	3	460	H/V/E
14.9	1925.8	454.2	10	3	208	H/V/E
14.9	1925.8	454.2	10	3	230	H/V/E
14.9	1925.8	454.2	10	3	460	H/V/E
18.6	2463.8	454.2	10	3	208	H/V/E
18.6	2463.8	454.2	10	3	230	H/V/E
18.6	2463.8	454.2	10	3	460	H/V/E
22.4	3313.4	454.2	9	3	208	H
22.4	3313.4	454.2	9	3	230	H
22.4	3313.4	454.2	9	3	460	H
29.8	5210.9	--	9	3	208	H
29.8	5210.9	--	9	3	230	H
29.8	5210.9	--	9	3	460	H
37.3	6230.4	--	9	3	208	H
37.3	6230.4	--	9	3	230	H
37.3	6230.4	--	9	3	460	H

H – Horizontal | V – Vertical | E - Enclosed

Specifications of 2-stage Air Compressors

Table A1.4 | Specifications of 2-stage compressors

kW	FAD lpm @6 bar	Tank size (litre)	Max. pressure (bar)	Phase (Electric supply)	Input voltage (VAC)	
3.7	396.5	302.8	12	3	200	V
3.7	396.5	302.8	12	3	200	V
3.7	396.5	302.8	12	1	230	V
3.7	396.5	302.8	12	3	230	V
3.7	396.5	302.8	12	1	230	V
3.7	396.5	302.8	12	3	230	V
3.7	396.5	302.8	12	3	460	V
3.7	396.5	302.8	12	3	460	V
3.7	470.1	302.8	12	3	208-230/ 460	H
3.7	470.1	302.8	12	1	208-230	H
3.7	470.1	302.8	12	1	208-230	V
3.7	470.1	302.8	12	3	208-230/ 460	V
5.6	679.7	302.8	12	3	200	V
5.6	679.7	302.8	12	3	230	V
5.6	679.7	302.8	12	3	230	V
5.6	679.7	302.8	12	3	460	V
5.6	688.2	302.8	12	3	208-230/ 460	H
5.6	688.2	302.8	12	1	208-230	V
5.6	688.2	302.8	12	3	208-230/ 460	V
7.5	965.7	454.2	12	3	208-230/ 460	H
7.5	965.7	454.2	12	3	208-230	H
7.5	965.7	340.7	12	3	200-208	V
7.5	965.7	340.7	12	3	208-230/ 460	V
7.5	991.2	454.2	12	3	200	H
7.5	991.2	454.2	12	3	230	H
7.5	991.2	454.2	12	3	460	H
7.5	991.2	454.2	12	3	200	V
7.5	991.2	454.2	12	3	230	V
7.5	991.2	454.2	12	3	460	V
11.2	1416.0	454.2	12	3	200	H
11.2	1416.0	454.2	12	3	230/460	H
11.2	1416.0	454.2	12	3	208-230/ 460	H
11.2	1416.0	454.2	12	3	200-208	H
14.9	1755.8	454.2	12	3	208-230/ 460	H
18.6	2378.9	454.2	12	3	208-230/ 460	H
22.4	2690.4	454.2	12	3	208-230/ 460	H

Specifications of 2-stage Pressure-lubricated Compressors

Table A1.5 | Specifications of 2-stage pressure-lubricated compressors

kW	FAD lpm @6 bar	Tank size (litre)	Max. pressure (bar)	Phase (Electric supply)	Input voltage (VAC)	
3.7	489.9	302.8	12	1	208-230	H
3.7	489.9	302.8	12	3	208-230/460	H
3.7	489.9	302.8	12	1	230	V
3.7	489.9	302.8	12	3	208-230/460	V
5.6	654.2	302.8	12	1	208-230	H
5.6	654.2	302.8	12	1	208-230	V
5.6	654.2	302.8	12	3	208-230/460	V
5.6	665.5	302.8	12	3	208-230/460	H
7.5	985.5	454.2	12	3	200-208	H
7.5	985.5	454.2	12	3	230/460	H
11.2	1387.7	454.2	12	3	230/460	H

H – Horizontal | V – Vertical

Appendix 2

General information on pneumatic actuators

1. Essential specifications of actuators

The essential technical specifications for pneumatic actuators are given in Table A2.1.

Table A2.1 | Typical specifications for pneumatic components:

1.	Medium	Compressed air, filtered, lubricated
2.	Operation	Doubling-acting, air-cushioned
3.	Operating pressure	0.1 to 12 bar
4.	Operating temperature	-10°C to +80°C
5.	Linear actuators:	
5.1	Size	Piston rod diameters from 1 mm to 320 mm
5.2	Thrust	2.7 N to 48000 N (at 6 bar)
5.3	Stroke length	1 mm to 10 m
5.4	Speed	5 to 15000 mm/s
6.	Rotary actuator:	
6.1	Size	Rotary drive diameter from 6 mm to 100 mm
6.2	Torque	0.15 Nm to 150 Nm (at 6 bar).
6.3.	Angle of rotation	1° to 360°
6.4.	Speed	Up to 50000 rpm

Annexure 3

Air Consumption Chart for Industrial Type Tools

Table A3.1 | Air consumption at 5 to 6 bar

Tool	Consumption (lpm) at 25% usage factor
Air motor, 1 hp	255
Air motor, 2 hp	510
Air motor, 3 hp	680
Burring tool, large	170
Burring tool, small	113
Chipping hammer	227
Die grinder, medium	170
Drill, 1/16" to 3/8"	170
Drill, 3/8" to 5/8"	255
Horizontal grinder, 2"	227
Horizontal grinder, 4"	425
Horizontal grinder, 6"	510
Horizontal grinder, 8"	566
Impact Wrench, 1"	312
Impact Wrench, 1/2", 5/8", 3/4"	227
Impact Wrench, 1/4"	113
Impact Wrench, 1¼"	396
Impact Wrench, 3/8"	142
Nut setters, large up to 3/4"	425
Nut setters, small up to 3/8"	170
Paint spray gun	142
Rammers, medium /large	283
Rammers, small	170
Riveting hammer, Heavy	227
Riveting hammer, Light	113
Saws, circular	453

Scaling hammer	85
Screw driver #2 to #6 screw	85
Screw driver #5 to 5/16" screw	170
Trapper, up to 3/8"	170
Vertical grinders and sanders 5" pad	255
Vertical grinders and sanders 7" / 9" pad	566

Note:

- *Air consumption is only indicative and may not be accurate for any particular make*
- *Always check with the OEM for the actual air consumption of tools*

Air Consumption Chart for Automotive Service Shops

Table A3.2 | Air consumption of tools for automotive service shops

Tool	Consumption lpm (FAD)	Max. Pressure (bar)
Portable Tools		
Blow Gun	71	6
Body Polisher	566	6
Body Sander	283	6
Brake Tester	113	6
Burring Tool	142	7
Die Grinder	142	6
Drill, 1/16" to 3/8"	113	6
Filing and Sawing Machine	142	7
Impact Wrench 3/8" to 1'"	85 - 340	6
Vertical Disc Sanders	566	7
TIRE TOOLS		
Air Hammer	113	7
Bead Breaker	340	10
Rim Stripper	170	10
Spring Oiler	113	7

Tire Hammer	340	7
Tire Inflation Line	57	10
Vacuum Cleaner	198	10
SPRAY GUNS		
Engine Cleaner	142	7
Paint Spray Gun (production)	227	7
Paint Spray Gun (touch up)	113	7
Paint Spray Gun (undercoat)	538	7
OTHER EQUIPMENT		
Car Lift, 8000 lb	170	12
Car Washer	255	7
Grease Gun	85	10
Medium Duty Sander	1133	7
Spark Plug Cleaner	142	7
Transmission Flusher	85	7

Fluid Power Educational Series Books

1. Pneumatic Systems and Circuits -Basic Level (In the SI Units)
2. Industrial Pneumatics -Basic Level (In the English Units)
3. Pneumatic Systems and Circuits -Advanced Level
4. Electro-Pneumatics and Automation
5. Design of Pneumatic Systems (In the SI Units)
6. Design Concepts in Pneumatic Systems (In the English Units)
7. Maintenance, Troubleshooting, and Safety in Pneumatic Systems
8. Industrial Hydraulic Systems and Circuits -Basic Level (In the SI Units)
9. Industrial Hydraulics -Basic Level (In the English Units)
10. Hydraulic Fluids
11. Hydraulic Filters: Construction, Installation Locations, and Specifications
12. Hydraulic Power Packs (In the SI Units)
13. Power Packs in Hydraulic Systems (In the English Units)
14. Hydraulic Cylinders (In the SI Units)
15. Hydraulic Linear Actuators (In the English Units)
16. Hydraulic Motors (In the SI Units)
17. Hydraulic Rotary Actuators (In the English Units)
18. Hydraulic Accumulators and Circuits (In the SI Units)
19. Accumulators in Hydraulic Systems (In the English Units)
20. Hydraulic Pipes, Tubes, and Hoses (In the SI Units)
21. Pipes, Tubes, and Hoses in Hydraulic Systems (In the English Units)
22. Design of Industrial Hydraulic Systems (In the SI Units)
23. Design Concepts in Industrial Hydraulic Systems (In the English Units)
24. Maintenance, Troubleshooting, and Safety in Hydraulic Systems
25. Hydrostatic Transmissions (HSTs) (In the SI Units)
26. Concepts of Hydrostatic Transmissions (In the English Units)
27. Load Sensing Hydraulic Systems (In the SI Units)
28. Concepts of Load Sensing Hydraulic Systems (In the English Units)
29. Electro-hydraulic Proportional Valves
30. Electro-hydraulic Servo Valves
31. Cartridge Valves
32. Electro-hydraulic Systems and Relay Circuits

For more details, please visit: **htpps://jojibooks.com**

About the Author

Joji Parambath is a trainer in the field of Pneumatics, Hydraulics, and PLC, for over 25 years. During his career, he has trained numerous professionals from the industries as well as faculty members and students of engineering institutions.

At present, he is the key trainer at Fluidsys Training Centre, Bangalore, India, (https://fluidsys.org) which is providing training in the field of Pneumatics and Hydraulics. He has already written two books on Pneumatics and Hydraulics. The publication of the present series of 32 books is intended to restructure and update the existing books.

The author wishes to thank all trainees for their lively interaction and many useful suggestions during the training programmes that prompted the author to write the present series of books. You may send your feedback to joji.p@hotmail.com

10th June 2020